United States High-Altitude Test Experiences: A Review Emphasizing the Impact on the Environment

Hermann Hoerlin

Contents

Publishing Information

Nimble Books LLC: The AI Lab for Book-Lovers

~ Fred Zimmerman, Editor ~

Humans and AI making books richer, more diverse, and more surprising.

(c) 2024 Nimble Books LLC

ISBN: 978-1-60888-308-0

AI-generated Keyword Phrases

- Starfish event radiation;
- High-altitude nuclear explosions;
- Air fluorescence;
- Upper atmosphere radiation changes;
- Teak and Orange nuclear tests;
- Natural radiation dose;
- Los Alamos Scientific Laboratory reports
- Space Research V publication;
- High-altitude explosions;
- Radiation effects at different altitudes;

Publisher's Notes

In a world grappling with the resurgence of nuclear threats and the ongoing exploration of space, this book serves as a vital reminder of the profound impact of human actions on the delicate balance of our planet and the cosmos. It compels us to confront the enduring questions surrounding the responsible use of technology, the pursuit of scientific knowledge, and the preservation of our environment for generations to come. Driven by a primal curiosity to unravel the mysteries of the universe and a deep-seated desire to safeguard our future, we are required to study the lessons of the past as we are compelled to face the complex challenges that lie ahead.

This annotated edition illustrates the capabilities of the AI Lab for Book-Lovers to add context and ease-of-use to manuscripts. It includes several types of abstracts, building from simplest to more complex: TLDR (one word), ELI5, TLDR (vanilla), Scientific Style, and Action Items; essays to increase viewpoint diversity, such as Grounds for Dissent, Red Team Critique, and MAGA Perspective; and Notable Passages and Nutshell Summaries for each page.

Abstracts

Analysis Based on Full Context

These analyses are created by using an LLM with a very long input context window, in this case Google Gemini 1.5-pro. The advantage is that the model can use the entirety of the document in its simulated reasoning.

This monograph provides a comprehensive review of the United States' high-altitude nuclear testing program conducted between 1955 and 1962, focusing on the environmental consequences of these tests. It meticulously details the chronology, locations, and yields of these explosions, while also delving into the complex phenomenology of weapon output interaction with the upper atmosphere. The study illuminates the formation of fireballs, energy partitioning, and debris distribution across vast distances, emphasizing the distinct differences observed at varying altitudes.

The monograph provides a meticulous examination of the impact of these high-altitude tests on various aspects of human activities and the environment. The author, drawing from a wealth of scientific observations and case studies, documents the effects on radio communications, satellite operations, and the creation of artificial radiation belts. The text analyzes the occurrence of phenomena like flash blindness, eyeburn hazards, and disruptions to communication networks, highlighting the severity of these impacts, especially following the high-yield Teak and Orange events.

Furthermore, the study addresses concerns regarding the potential effects of high-altitude explosions on weather patterns, the depletion of the ozone layer, and the persistence of radioactive tracers in the stratosphere. While acknowledging the apprehension that initially surrounded these tests, the author concludes by highlighting the substantial scientific insights gained from these high-altitude experiments. He argues that these tests, while not without their drawbacks, ultimately contributed significantly to our understanding of atmospheric physics, plasma dynamics, and the complex interplay between nuclear detonations and the Earth's magnetosphere.

Analysis Based on Abridged Content Windows

These analyzes are created by analyzing portions or summaries of the document, using LLMs with shorter context windows. The advantage is that these models are faster and cheaper.

TL;DR (one word)

Radiation.

Explain It To Me Like I'm Five Years Old

Imagine a big explosion happened really high up in the sky, about 400 kilometers above the Earth. This explosion caused a lot of radiation, which is like invisible energy that can be harmful. This radiation can affect things like the air and the atmosphere around us. Scientists have studied these explosions to learn more about how they can impact our planet.

TL;DR (vanilla)

The Starfish event was a high-altitude nuclear explosion that occurred 400 km above the Earth's surface. It resulted in a significant increase in radiation levels, about 1000 times the natural dose. This event, along with others, contributed to changes in the upper atmosphere and increased levels of air fluorescence.

Scientific Style

The Starfish event, a high-altitude nuclear explosion occurring 400 km above the Earth's surface, resulted in drastic changes in radiation levels in the upper atmosphere. This event, which occurred 40 years ago, produced a radiation dose approximately 1000 times higher than natural levels. Research conducted by Hoerlin and Galperin et al. has shed light on the effects of high-altitude nuclear explosions on air fluorescence and radiation levels. These findings highlight the potential impact of such events on the Earth's upper atmosphere and emphasize the need for further research in this area.

Action Items

Further research on the effects of high-altitude nuclear explosions on the upper atmosphere and radiation levels.

Monitoring and studying the long-term effects of these events on the environment and human health.

Implementing measures to prevent and regulate high-altitude nuclear explosions to minimize potential harm.

Educating the public and policymakers on the risks and consequences of such events.

Viewpoints

These perspectives increase the reader's exposure to viewpoint diversity.

Grounds for Dissent

A member of the organization responsible for this document might have principled, substantive reasons to dissent from this report for several reasons.

Ethical Concerns: The member may have ethical concerns about the use of high-altitude nuclear explosions and the resulting radiation exposure. They may believe that the potential harm to human health and the environment outweigh any potential benefits of such actions.

Scientific Disagreement: The member may have scientific disagreement with the conclusions drawn in the report. They may believe that the data presented is incomplete or that the methodology used is flawed, leading to inaccurate or misleading results.

Policy Differences: The member may have policy differences with the organization's stance on high-altitude nuclear explosions. They may believe that alternative approaches should be considered or that more research is needed before making any decisions on this issue.

Public Safety Concerns: The member may have concerns about the potential impact of high-altitude nuclear explosions on public safety. They may believe that the risks of such actions are too great and that more precautions should be taken to protect the public from harm.

Overall, the member's dissenting views may be rooted in a commitment to ethical principles, scientific integrity, policy considerations, and public safety concerns. They may believe that the organization's stance on high-altitude nuclear explosions is misguided and that alternative approaches should be pursued.

Red Team Critique

Lack of Clarity: The document is unclear and lacks context. It jumps between different altitudes and events without providing a clear explanation of their significance or relevance. Additionally, the reference to the Starfish event occurring at 400 km above JI is confusing and needs further explanation.

Lack of Analysis: The document provides factual information about radiation levels and events, but it lacks in-depth analysis or interpretation of the data. It fails to discuss the potential implications or consequences of high-altitude nuclear explosions on the upper atmosphere.

Lack of Sources: The document references a report from the Los Alamos Scientific Laboratory and a study by Galperin and Bolyunova, but it does not provide any additional sources or evidence to support its claims. Without additional sources, the credibility of the information presented is questionable.

Inconsistencies: The document mentions a radiation dose that is about 1000 times the natural dose, but it does not explain how this calculation was made or what the significance of this information is. The inconsistency in providing relevant details makes it difficult to fully understand the impact of high-altitude nuclear explosions on radiation levels.

Lack of Conclusion: The document ends abruptly without a clear conclusion or summary of the key points discussed. It would benefit from a concluding section that ties together the information presented and highlights the main findings or implications of the research.

Overall, the document lacks depth, clarity, and coherence. It would benefit from a more structured and analytical approach to discussing the effects of high-altitude nuclear explosions on the upper atmosphere. Additionally, providing more detailed explanations, sources, and analysis would enhance the credibility and impact of the document.

MAGA Perspective

This document is just another example of fear-mongering and propaganda from the left to demonize the MAGA movement. The mention of high-altitude nuclear explosions is clearly an attempt to paint the Trump administration as reckless and dangerous. This is nothing more than an attempt to discredit our President and his policies.

The so-called "Starfish event" mentioned in this document is being used to create hysteria and panic among the American people. The suggestion that this explosion occurred at 400 km above JI is absurd and baseless. This is just another attempt by the mainstream media to undermine the credibility of MAGA supporters.

The idea that this event resulted in a radiation dose 1000 times the natural level is completely unfounded and lacks any scientific evidence to back it up. This is just another example of the left using scare tactics to push their socialist agenda.

The mention of the Starfish explosion occurring 400 km above JI is clearly a fabrication designed to create panic and fear. This kind of misinformation is exactly why the American people have lost trust in the mainstream media and the political establishment.

The reference to high-altitude explosions being responsible for a fraction of the radiation in the upper atmosphere is clearly an attempt to blame the Trump administration for something that is beyond their control. This document is nothing more than political propaganda meant to distract the American people from the real issues facing our country.

Page-by-Page Summaries

BODY-1 Review of high-altitude test experiences at the University of California, emphasizing the impact on the environment.

BODY-2 Information about work performed under a contract with the US Energy Research and Development Administration, available for purchase from the National Technical Information Service. No warranty or liability is assumed for the accuracy or completeness of the information provided.

BODY-3 Review of high-altitude test experiences in the US, focusing on environmental impact.

BODY-5 The page discusses various effects of thermal radiation, flash blindness, and electromagnetic radiation on different systems and phenomena, including satellites, communication, and weather patterns. It also addresses concerns about environmental effects of nuclear explosions in space.

BODY-6 An image of an orange event seen from a US aircraft camera at approximately 1 minute after the event.

BODY-7 Review of US high-altitude nuclear explosions from 1955-1962, detailing impacts on environment, atmosphere, and human activities. Observations on thermal effects, ionospheric ionization, and artificial radiation belts. Insights gained outweigh damages sustained. Potential effects on manned space flight considered.

BODY-8 High-altitude explosions led to spectacular auroral phenomena, increased knowledge of light-producing processes, and minimal fallout effects on humans. Tracers helped understand air-mass motions and mixing processes. Overall, effects on populations were insignificant, but potential consequences of massive military operations in the upper atmosphere are grave.

BODY-9 The chapter provides a listing of high-altitude events in the US, describing the differences in explosion phenomena between low and high-altitude events, including thermal damage, radiative expansion, and worldwide motion of debris.

BODY-11 High-altitude nuclear tests, such as Teak and Kingfish, had significant effects on communication and debris motion due to magnetic field interactions. Argus experiments produced artificial radiation belts, while Starfish created long-lived belts interfering with natural phenomena. Thermal output was minimal, with less severe communication interference compared to lower-altitude tests.

BODY-12 Aerial views of Bluegill and Teak events from different perspectives.

BODY-13 Aerial views of Kingfish and Checkmate events from high-flying aircraft and Johnston Island.

BODY-14 Images of the Starfish event, showing air fluorescence excited by debris particles and magnetic field lines, captured by cameras on aircraft and Christmas

Island.

BODY-15 Prompt thermal radiation is emitted by fireballs in a few seconds, causing damage and injury. Altitude affects radiation output, with lower output at higher altitudes. High-altitude explosions had minimal environmental impact, except for blinding birds at Teak and slight sunburn on a person.

BODY-16 Table listing approximate thermal doses from high-altitude explosions, with significant magnitude only for Teak and Orange events. Difficulty in predicting thermal flux directly under shots reported.

BODY-17 Flash blindness is a temporary loss of vision from exposure to high-intensity light sources, with recovery time depending on various factors. Eyeburn, on the other hand, involves permanent damage to the retina from focused light sources, with potential for irreversible damage and blind spots.

BODY-18 The page discusses the history, studies, and safe dosage levels for protecting the eyes from nuclear-burst eyeburn problems, with a focus on factors such as image diameter, exposure times, and source temperatures.

BODY-19 The page discusses the eye-damage hazard from nuclear test explosions at different altitudes, highlighting specific cases of eyeburn injuries and the importance of protective measures.

BODY-20 Table showing retinal dose at Ground Zero, safe dose, and approximate safe slant distance for various events. Case studies of two individuals with chorioretinal burns near the fovea after exposure to radiation, showing varying levels of visual impairment and impact on daily life.

BODY-21 Discussion of the Teak eyeburn hazard problem, including theoretical calculations, potential damage to the retina, and decision to move events to unpopulated area due to danger to Marshall Island natives.

BODY-22 Research on the effects of nuclear explosions on primates led to a better understanding of safe exposure levels, with irreversible damage occurring at a 20°C temperature increase and a 5°C rise being safe. This research was applied to the Teak case, reducing the safe dose levels.

BODY-23 High-altitude events caused ionization leading to radio communication degradation in the Pacific. Effects varied based on fission yield and debris location. Teak and Orange events had the most severe impact, while Starfish had different interference patterns due to burst altitude differences.

BODY-24 Effects of nuclear tests Teak, Orange, and Starfish on communication systems in the Pacific were severe, causing blackouts and disruptions in LF, MF, and HF propagation. VHF transmissions improved due to increased electron densities. Limited information on other events due to classified data.

BODY-25 Effects of Teak and Orange nuclear tests on radio communication in the South Pacific, including blackouts and enhanced signal propagation. Notable changes in signal strength over HF links from Japan to Honolulu and San Francisco.

BODY-26 Effects of the Starfish Prime nuclear test on Johnston Island and surrounding areas caused significant communication disruptions, with signal strength drops and blackout periods observed in New Zealand, Australia, and other locations. The impact extended to distant regions, highlighting the widespread consequences of high-altitude nuclear tests.

BODY-27 HF interference caused communication disruptions in the JI and Pacific areas for several hours. Ionosonde measurements showed blackout of HF frequencies due to heavy absorption in the D-layer. Effects were moderate compared to other events, with minimal worldwide geomagnetic effects. Delayed interference in Hawaii area after Teak and Orange events due to gamma and beta-ray flux.

BODY-28 Incomplete report on communication interference during atmospheric interactions, with limited observational data hindering a full understanding of the phenomenon.

BODY-29 Artificial radiation belts were created in the South Atlantic through nuclear explosions in Project Argus, forming trapped radiation similar to the natural Van Allen belts. These artificial belts were stable for weeks and had higher flux densities than the natural belts.

BODY-30 The page discusses the impact of nuclear tests on the Earth's radiation belts, focusing on events such as Teak, Orange, and Starfish. It highlights the formation of artificial radiation belts and the scientific observations made from these events.

BODY-31 The page discusses the injection of electrons into the radiation belt after the Starfish Prime nuclear test, their decay over time, the contribution of neutrons, and the lack of evidence for the formation of artificial radiation belts. It also mentions satellite damage from the event.

BODY-32 Satellite Traac experienced solar cell degradation in space before failing to transmit data after an explosion. Different satellites faced varying issues and lifetimes due to radiation exposure, highlighting the complexity of space travel effects on spacecraft and astronauts.

BODY-33 Radiation exposure limits for space explorers, Gemini spacecraft shielding details, Russian high-altitude tests in 1962, and mixed scientific community response.

BODY-34 A summary of the page content in less than 50 words.

BODY-35 Synchrotrons radiation and hydromagnetic waves were observed after the Starfish explosion, with radiation detected in various locations and magnetic disturbances noted. The radiation intensity decayed slowly, and the disturbances were associated with the stretching of magnetic field lines.

BODY-36 Magnetic field disturbances caused by nuclear radiation interactions with the ionosphere led to increased cosmic-ray flux in Peru, with various interpretations provided by different researchers. The disturbances did not interfere with geophysical station activities but contributed to understanding natural phenomena.

BODY-37 Electromagnetic radiation from high-altitude detonations can cause problems in electronic systems, with potential failures in receivers and rectifiers. Solid-state circuitry increases susceptibility, but problems are minimized if detonations are below line-of-sight.

BODY-39 Detection of nuclear explosions in space can produce auroral phenomena similar to natural auroras, with brilliant fluorescence emissions at higher altitudes. The Teak event provided a breathtaking display, while the Starfish aurorae were less brilliant but covered a larger area.

BODY-40 A study on the effects of nuclear explosions in space, including auroral observations and potential weather pattern impacts, with a focus on the Starfish event and detection systems.

BODY-41 The page discusses the relationship between magnetic storms, vorticity area index, auroral energy depositions, and potential weather modification through nuclear explosions, highlighting the complex interactions between the Earth's atmosphere and external factors.

BODY-43 Radioactive tracers injected into the stratosphere from nuclear explosions showed movement patterns and distribution in both hemispheres, with faster vertical motion at high latitudes. Initial observations in the south were followed by similar concentrations in the north, indicating movement towards the winter hemisphere.

BODY-44 The page discusses the global inventory of radioactive debris in the atmosphere, with observations confirming results suggested by tracers. It also delves into the concept of residence time for radioactive debris in the stratosphere and troposphere, highlighting differences based on injection altitudes and explosion types.

BODY-45 Debris injected into the stratosphere has residence times of 1-2 years, with downward and equatorward movement. Studies using tracers like '02Rh show summer-to-winter hemisphere flow and descending motion in the winter stratosphere. Models can predict fallout parameters and deposition of debris in the atmosphere.

BODY-46 The page discusses the impact of radioactive tracer data on global atmospheric and stratospheric physics, noting a decrease in relevant publications after 1972. It highlights the use of carbon-14 data from nuclear explosions to study air mass exchange between troposphere and stratosphere.

BODY-47 Fission-produced bromine and iodine from nuclear explosions have minimal impact on ozone layer, with insignificant effects from stratospheric injection. Local fallout from high-altitude events showed no significant activity.

BODY-49 Debate over whether nuclear tests would damage ozone layer, calculations show minimal impact, high-altitude tests had little effect, Russian tests had more significant impact, precise effects still debated.

BODY-51 Sir Bernard Lovell expressed concern about the environmental effects of starfish-type explosions in 1962, urging international scientific agreement to prevent militarization of space and destruction of astronomy on Earth.

BODY-52 International agreement on space use is urgent. Concerns raised by NASA staff about artificial radiation belts interfering with space program. Emotional response led to cancellation of nuclear space experiments. Despite initial setbacks, high-altitude events led to scientific advancements and stimulated research in various fields.

BODY-53 Research on high-altitude explosions led to insights on geophysical processes, the need for precise reaction rates, and advancements in atmospheric research and plasma physics.

BODY-54 Acknowledgments for work done under the Nevada Operations Office of the US Energy Research and Development Administration, with thanks to various individuals for their assistance and contributions to the project.

BODY-55 References on ocular effects of nuclear detonations and related topics, including flash blindness, retinal burns, and safe separation distances.

BODY-56 Various scientific reports and communications detailing the effects of high-altitude nuclear detonations on human health, radio communications, ionospheric disturbances, and geophysical phenomena.

BODY-57 Research papers and reports on the effects of artificially injected electrons into the geomagnetic field, trapped radiation from nuclear devices, and the artificial radiation belt from a nuclear detonation in 1962.

BODY-58 Research articles on artificial radiation belts, energetic particles from nuclear explosions, and radiation hazards in space flight are discussed in this page.

BODY-59 Various studies on the effects of high-altitude nuclear explosions on the atmosphere and geomagnetic field, including observations of synchrotron radiation and fluorescence of air excited by energetic electrons and X-rays.

BODY-60 Research on nuclear explosions in space, air fluorescence detection systems, solar magnetic sector structure effects on weather, and modifying the ionosphere with radio waves and gas releases.

BODY-61 Various reports and studies on the use of radioactive tracers, such as cadmium isotopes and strontium-90, to investigate atmospheric circulation and fallout patterns from nuclear testing.

BODY-62 Various reports and studies on atmospheric nuclear detonations, stratospheric tracer concentration, atmospheric transport models, and effects of atomic radiation.

BODY-63 A list of scientific publications and reports related to space research, ionizing radiation, and high-altitude nuclear explosions.

Notable Passages

BODY-7 The formation of an artificial radiation belt of such high electron fluxes and long lifetimes as occurred after the Starfish event was unexpected; so were the damages sustained by three satellites in orbit. However, the vast amount of knowledge gained by the observations of the artificial belts generated by Starfish, Argus, and the Russian high-altitude explosions far outweighed the information which would have been gained otherwise.

BODY-8 The worldwide auroral phenomena produced by the high-yield explosions were spectacular but of no consequence to ordinary human activities. They increased substantially our basic knowledge of auroral-type light-producing processes. Questions were raised but not answered as to the effects of pertinent energy depositions on large-scale weather patterns. The prompt fallout from high-altitude explosions was zero. The residence time in the stratosphere of special tracers—'02Rh and'"'Cd—incorporated into the Orange and Starfish devices was 14 years. The fallout of fission products was similarly delayed and was distributed over the whole globe; thus, the biological effects on humans were reduced per unit energy release in comparison with low-altitude atmospheric explosions.

BODY-9 The explosion phenomena of the three "low" high-altitude, low-yield events (HA, John, and Yucca) did not differ drastically from that of sea-level or near-sea-level explosions. Compared with the latter, the power-time histories of the fireballs were somewhat shorter, the peak radiances were slightly higher, the time of the minimum (bhangmeter time) was shorter, and the thermal pulse was more intense. While the total prompt-thermal-yield fraction was close to the thermal fraction of similar events near sea level, the somewhat shorter duration of the pulse is more effective in producing thermal damage. Therefore, the thermal-radiation effects of explosions in the "low" high-altitude domain would be more severe

BODY-11 The events produced the first artificial radiation belts, shortly after the discovery by Van Allen of the natural belts. Finally, Starfish was fired with a yield of 1.4 Mt at 400 km altitude above Johnston Island (JI) (next to Argus II the highest event), While many of the results were of military value, Starfish was also an experiment of worldwide scientific interest. Yield, altitude, and time of event were announced prior to the event. At Starfish altitude, magnetic pressure and air-particle pressure are of about the same magnitude; therefore, the field effects play a very strong role from the earliest time on in the event. Indeed, the debris motion was largely governed by the magnetic field.

BODY-15 Prompt thermal radiation is defined as that part of the fireball radiation which is emitted in times of the order of a few seconds or less. This radiation, A >3200 ~, is transmitted by undisturbed air to long distances. If intense enough, it can produce damage to materials and serious injury to humans. The power peak of the radiation at sea level occurs roughly at the time of the second maximum: t2nd III,,= 0.03 flseconds, where Y is in kilotons. At 10 times second maximum time, the rate of emission has decreased by one or several powers of 10.

BODY-16 ".. .we have held several discussions here among UCRL (Livermore) personnel.. and also Mike May and Tom Wainwright have discussed this problem with Al Latter and some of his people at Rand. This group has been unable to devise a model which we agree could be relied upon with any degree of confidence to predict the thermal flux on the ground directly under this shot."

BODY-17 Flash blindness is defined as the temporary loss of vision resulting from exposure to a high-intensity light source from which an after-image develops. An extensive literature, including results from several laboratory tests, describes a large variety of exposure conditions. The recovery time depends on numerous variables such as flash luminance, flash duration, source spectrum and geometry, flash distance, and degree of eye-adaptation of the observer. The regeneration of the visual pigments and an apparent automatic brightness control which reduces the sensitivity of the bleached retinal area are age-dependent; there are also wide variations in responses from subject to subject.

BODY-18 Predictions of the early radiance of the fireball yielded numbers much in excess of the radiance of the solar photosphere. Consequently, observers used high-density dark goggles for protection. Use of goggles with density 4, attenuating the light fluxes by a factor of 10000, became the rule in subsequent tests. In the 1950's, semiquantitative studies were made during several US test operations. More quantitative investigations were made in several laboratories, such as at the Medical college of Virginia (Dr. W. T. Ham and associates) and the Ophthalmology Department of the United States Air Force School of Aerospace Medicine, Brooks Air Force Base, Texas, and others.

BODY-19 "At still higher altitudes, as with the Starfish event (400 km above JI), the fireball phenomenology is changed again: the main thermal x-ray energy radiated by the source has a very long mfp and is absorbed over a very large volume of air at about 100 km altitude, thus producing mainly air fluorescence, little heating of the air, and relatively low radiance—i.e., low optical power per unit area, with no hazard from this source."

BODY-20 "In a recent review of these two cases, the fact that chorioretinal burns on or near the fovea do not necessarily cause complete blindness was emphasized. Both size and location of the lesion determine visual impairment."

BODY-21 "The eyeburn hazard was considered to be serious. The consensus was that further study was needed. It was felt that dark goggles were certainly needed at Rongerik, an inhabited atoll 250 km from Bikini. The seriousness of the problem was subsequently relayed to the Eniwetok Planning Board meeting at NTS on May 14, 1957, by Ogle."

BODY-22 We believe now that irreversible damage occurs for a temperature increase of 20"C, while a 5°C temperature rise is safe. Applying these criteria to the Teak case, the threshold (20"C) dose at ground zero would reduce from 3 cal/cm2 to 1 cal/cm2 on the retina and the safe dose to 0.2 cal/cm'.

BODY-23 The ionization was caused mainly, but not exclusively, by fission-product gamma rays, and, in more local areas, by beta radiation. The extent and intensity of ionization was governed by the location of the debris and by the fission yield. When the debris rise from the burst location to higher altitudes, they spread in space, and the gamma rays, because of their long mfp, cover increasingly larger geographic areas. The debris cloud as such is also highly ionized because of the short range of the beta particles, although many betas escape and produce ionization in conjugate areas.

BODY-24 Later, the debris patches were observed to rise from the conjugate areas, and to spread over still larger areas of the Pacific. Thus, the Starfish debris space-time history differed significantly from the Teak and Orange histories. While D-layer absorption of radio frequencies was the main cause of the communication blackout, it is interesting to note that long-distance VHF transmissions improved after all three events in several areas, particularly at night, because of the increased electron densities in the E- and F-regions.

BODY-25 "However, 'phenomenally good high frequency communication' became possible because of the abnormally high ionization density in the F-layer. Signals on 30 MHz and above were heard over long distances even at night."

BODY-26 "In the HF range, total blackout occurred on JI only for a short time; moderate interference lasted for several hours. Absorption increased at sunrise. The debris patch in the northern conjugate area should have affected communications in the French Frigate Shoal, Midway, and Wake areas; also on board ship—the "DAMP" ship, for instance."

BODY-27 There is little information in the unclassified literature, presumably because yield and altitude are still classified. Interpretation of observations is therefore ambiguous. Degradation of communications was relatively moderate (compared with Teak, Orange, and Starfish). Geomagnetic worldwide effects were, by many orders of magnitude, smaller than for Starfish-understandably so, because at low altitude, the particle pressure is the dominant factor in fireball phenomenology. Very moderate southern conjugate ionospheric effects occurred. It is worthwhile to have a sharper look at the reasons why the communications interference in the Hawaii area was delayed after the Teak event and more so after Orange.

BODY-28 It would be desirable to present a still better, fully coherent story of the whole pertinent phenomenology.

BODY-29 Before the discovery of the natural Van Allen belts in 1958, N. C. Christofilos had suggested in October 1957 that many observable geophysical effects could be produced by a nuclear explosion at high altitude in the upper atmosphere. This suggestion was reduced to practice with the sponsorship of the Advanced Research Project Agency (ARPA) of the Department of Defense and under the overall direction of Herbert York, who was then Chief Scientist of

ARPA.

BODY-30 The Starfish event produced by far the most intense, long-lasting radiation belt. At burst time, several satellites were in rather low orbits; their apogees were near 1000 km. At D + 1 day, Telstar was launched into a more favorable elliptical orbit which covered the space up to an apogee of 5600 km. Subsequently, several other satellites were launched which provided additional data. For a listing, see the "Trapped Radiation Handbook." Sections 6-33. Because of differences in orbits, spectral coverage, and launch times, the data obtained by the various satellites did not always agree. However, the maximum electron fluxes were encountered between $L = 1.2$ and $L = 1$

BODY-31 "A probable mean number is 4×10^{27}, corresponding to an injection efficiency of -5%. It is most likely that the injection occurred by way of the strong debris-jets moving across the magnetic field lines as observed from Christmas Island and by a high flying aircraft."

BODY-32 "The vulnerability of solar cells and electronic circuit components to nuclear radiations has been treated extensively in the literature. Effects on Manned Spacecraft This is a complicated subject; the dose received by an astronaut depends on many variables such as type of orbit (ap~gee, perigee, inclination), degree of shielding, and duration of flights."

BODY-33 "The response was mixed."

BODY-35 "Electrons moving in a circular orbit perpendicular to a magnetic field will be accelerated and, consequently, emit radiation. Low-energy electrons radiate at the frequency of the circular motion; this frequency is often called the cyclotron frequency. High-energy relativistic electrons emit radiation in the direction of the motion of the particle at frequencies higher than the cyclotron frequency. This is called synchrotrons radiation or, sometimes, magnetic bremsstrahlung."

BODY-36 The coincidence of geomagnetic field fluctuations and variations in E-layer ionization density is advanced by another group as evidence that the perturbations were caused by interactions of nuclear radiations with the magnetic field and ionosphere.

BODY-37 The resulting transverse current in the large area of gamma-ray deposition produces a large coherent radiating element. With appropriate yield, detonation altitude, and magnetic azimuth, the electric fields over large areas at the earth's surface can exceed 104V/m. Such fields can cause detrimental effects on some types of electrical systems. The pulse width is less than a microsecond. Starfish produced the largest fields of the high-altitude detonations; they caused outages of the series-connected street-lighting systems of Oahu (Hawaii), probable failure of a microwave repeating station on Kauai, failure of the input stages of ionospheric sounders and damage to rectifiers in communication receivers.

BODY-39 "Gamma-ray-excited air fluorescence was already observed at Trinity. Most of the radiation is emitted from excited nitrogen, more specifically by the second and first positive systems of N2 and the first negative and the Meinel systems of N2+. The efficiencies for conversion of nuclear emissions into light is low at sea level because of collisional deactivation of the excited states; it is much higher at lower air densities. Consequently, at higher altitudes, all fluorescence emissions are more brilliant and quite spectacular indeed."

BODY-40 The most interesting result of this survey is the very large extent of the affected area: debris and debris electrons deposited their energies over some 40° of latitude in the south (Tarawa observed the luminosity in the overhead tube). These observations supplement the air-based gamma-ray measurements reported by D'Arty and Colgate. These auroral observations, together with others taken from airplanes, from Hawaii, and from Christmas Island, were invaluable in deriving a general physical picture of the Starfish phenomenology. Starfish was indeed a large-scale demonstration of many principles of plasma and auroral physics. The results confirmed the anticipation of the scientific usefulness of nuclear explosions in space as expressed in 1959 by LASL staff.

BODY-41 "The study of the coupling processes between the thermosphere where auroral particles and x-rays are stopped and the mesosphere, stratosphere, and the upper troposphere—i.e., the meteorologically important 300-millibar, -9-km-altitude range, remains a most interesting unsolved problem of upper-atmospheric physics."

BODY-43 The Orange and Starfish warheads contained special tracer elements created by neutron activation from the devices. About 3 megacuries (MCi) of 102Rh were produced in the Orange weapon; this isomer of principal concern has a half-life of 210 days. The main debris mass rose to an estimated altitude of 150 km, although relatively crude observations from Mt. Haleakala on Maui Island indicated some debris as high as 500 km. In the Starfish tracer experiment, 0.25 + 0.15 MCi of 109Cd were produced. Cadmium-109 has a half-life of 470 days. The Starfish explosion occurred 400 km above JI.

BODY-44 "Cadmium-109 was first observed by the AEC's balloon-sampling program in December 1962 at 35 km altitude in the south and several months later in the north. On the whole, the observations confirmed the results suggested by the 1°2Rh tracer motion; but in this case, the concentrations measured at high latitudes both in the north and the south were up to 10 times higher than in the equatorial areas."

BODY-45 "The tracer data indicate a summer-to-winter hemisphere flow above about 37 km and a mean descending motion in the winter stratosphere between 25° and about 70°. Ascending motion occurs near the equatorial tropopause and in the lower winter stratosphere poleward of 70°. Virtually the entire summer stratosphere and the winter stratosphere equatorward of 25° between 18 and 25 km is dominated by mixing processes with no evidence of organized circulations in the meridional plane."

BODY-46 "It remains to be seen whether or not observations of radioactive tracers and their interpretations have made an impact on the science of global atmospheric and stratospheric physics."

BODY-47 "In view of the current controversy on the effects of halogen gases on natural ozone and because of highly confused recent press reports which claim that injection of a few kilograms of bromine gas into the stratosphere would seriously affect the ozone concentration, John Zinn and I looked into the matter of bromine production by fission."

BODY-49 After the events, little attention was paid to this particular problem, evidently because no spectacular or unusual observations were made (because of lack of evidence one way or the other).

BODY-51 "May I conclude by saying that in spite of this enthusiasm which I display and this optimism for the future of scientific research I must confess that my belief in the inevitability of progress has been very considerably undermined during this past year by the realization that some of the American and probably some of the Russian space activities are not being guided by the purest of scientific motives."

BODY-52 "The need for international agreement about the use of space and the control of launchings, either of rockets or space-vehicles into it, has become a matter of the utmost urgency."

BODY-53 "The many deviations from equilibrium which are characteristic of high-altitude shots make it very difficult to make definite predictions on either hydrodynamic or optical phenomena...All I can hope to do is indicate the scale of the phenomena, not the details. This makes it more interesting to make observations on this test. In fact, the test will constitute a beautiful laboratory for the study of the properties of air in large quantity and at very low density."

LA-6405
c.3

UC-2 and UC-11
Reporting Date: June 1976
Issued: October 1976

UNITED STATES HIGH-ALTITUDE TEST EXPERIENCES

A Review Emphasizing the Impact

on the Environment

A LASL MONOGRAPH

by

Herman Hoerlin

los alamos
scientific laboratory
of the University of California
LOS ALAMOS, NEW MEXICO 87545

An Affirmative Action/Equal Opportunity Employer

UNITED STATES
ENERGY RESEARCH AND DEVELOPMENT ADMINISTRATION
CONTRACT W-7405-ENG. 36

Work performed under Contract No. At(26-1)-648 with the US Energy
Research and Development Administration, Nevada Operations Office,
Las Vegas, Nevada, 89114.

LA-6405

UC-2 and UC-11
Reporting Date: June 1976
Issued: October 1976

UNITED STATES HIGH-ALTITUDE TEST EXPERIENCES

A Review Emphasizing the Impact

on the Environment

A LASL MONOGRAPH

by

Herman Hoerlin

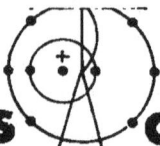

los alamos
scientific laboratory
of the University of California
LOS ALAMOS, NEW MEXICO 87545

An Affirmative Action/Equal Opportunity Employer

UNITED STATES
ENERGY RESEARCH AND DEVELOPMENT ADMINISTRATION
CONTRACT W-7405-ENG. 36

BODY-3

CONTENTS

v

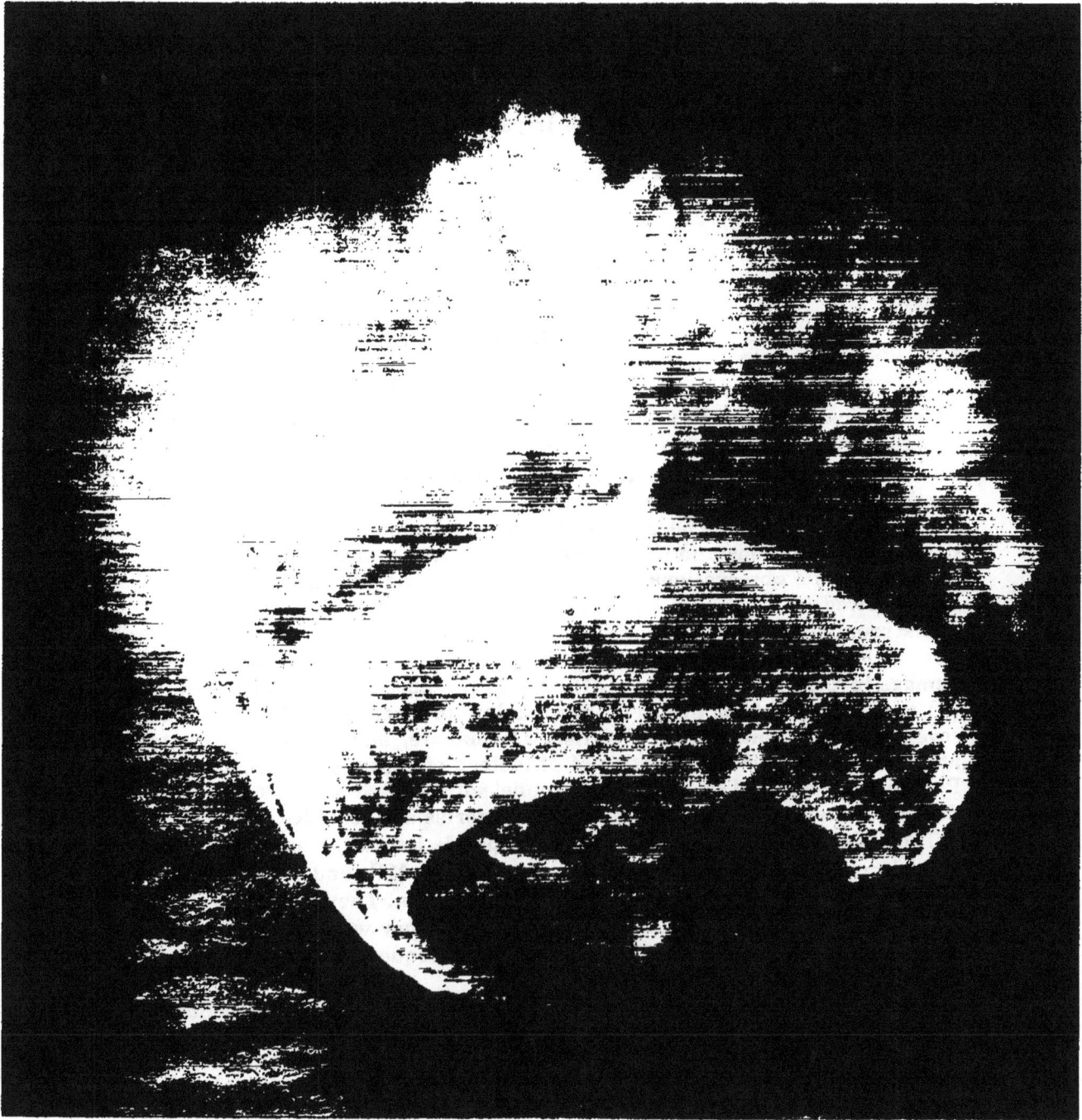

Fig. 1.
Orange Event seen from US aircraft carrier at approximately H + 1 minute.

UNITED STATES HIGH-ALTITUDE
TEST EXPERIENCES

A Review Emphasizing the Impact on the Environment

A LASL Monograph

by

Herman Hoerlin

ABSTRACT

The US high-altitude nuclear explosions of the 1955-1962 period are listed chronologically; dates, locations, and yields are given. The major physical phases of the interactions of the weapon outputs with the atmosphere are described, such as the formation of fireballs at the low high-altitudes and the partition of energies and their distribution over very large spaces at the higher high-altitudes. The effects of these explosions on the normal activities of populations and the protective measures taken are documented. Many scientific observations, together with their significance and values, are reviewed.

The prompt thermal effects on the ground were negligible, with the exception of those from the Orange event. That event could have caused minor damage in the Johnston Island (JI) area in the absence of cloud cover.

The eyeburn problem at ground zero and up to large slant distances was severe for all events except Starfish, Checkmate, and Argus. Adequate precautions, such as the selection of JI instead of Bikini as the base in the Pacific, were taken. Two military personnel suffered severe burns, however, due to inadvertent exposure. Their case histories are recorded.

The degrading effects of increased ionospheric ionization on commercial and aircraft communications—mainly in the LF, MF, and HF frequency ranges—extended over the whole Pacific Ocean area. They lasted for many days after the three megaton-range explosions. They were less severe—in some cases even beneficial—for VHF and VLF frequencies, thus providing guidance for emergency situations.

The formation of an artificial radiation belt of such high electron fluxes and long lifetimes as occurred after the Starfish event was unexpected; so were the damages sustained by three satellites in orbit. However, the vast amount of knowledge gained by the observations of the artificial belts generated by Starfish, Argus, and the Russian high-altitude explosions far outweighed the information which would have been gained otherwise. A few extrapolations are made to effects on manned space flight under hypothetical circumstances.

Electromagnetic radiation in the radio-frequency portion of the spectrum (EMP) caused brief outages of a street lighting system in Oahu and of several input stages of electronic equipment, though during the Starfish event only.

1

The worldwide auroral phenomena produced by the high-yield explosions were spectacular but of no consequence to ordinary human activities. They increased substantially our basic knowledge of auroral-type light-producing processes. Questions were raised but not answered as to the effects of pertinent energy depositions on large-scale weather patterns.

The prompt fallout from high-altitude explosions was zero. The residence time in the stratosphere of special tracers—^{102}Rh and ^{109}Cd—incorporated into the Orange and Starfish devices was 14 years. The fallout of fission products was similarly delayed and was distributed over the whole globe; thus, the biological effects on humans were reduced per unit energy release in comparison with low-altitude atmospheric explosions. The worldwide observation of the tracers led to the development of matching models of global stratospheric air-mass motions and to a better understanding of mixing processes near the tropopause. In fact, the downward motion of the tracers was most pronounced in the polar areas during local winter. No effect on the natural ozone layer could be ascertained.

In summary, the effects of the US high-altitude explosions on the normal activities of the populations were either insignificant or under protective control involving little harassment or irritation. As to the effects on the research activities of the international scientific community, I believe, in retrospect, that the early apprehension both in the US and Great Britain has given way now to a more positive assessment of the scientific returns obtained. However, it is also evident that the consequences of *massive* military operations in the upper atmosphere would be grave.

——————————————————————————

2

CHAPTER I

LISTING OF EVENTS.
GENERAL DESCRIPTION OF PHENOMENOLOGY.

All US high-altitude events are listed in Table I in temporal sequence. Much, but not all, of the information is taken from Glasstone.[1] The explosion times are rounded to the nearest minute, which is adequate for the purposes of this paper. The numbers were checked and supplemented by data from other sources. There is some uncertainty as to the Argus burst locations. They are different from the Glasstone data. The references used are shown at the bottom of Table I.

For the purpose of introduction, a general description of the main aspects of the phenomenology of the events is provided first. Later in the document, those phases of the phenomenology which are pertinent to a specific environmental effect are described in greater detail. The treatment will not always be entirely satisfactory to the pure scientist: it suffers from the non-utilization of precise weapons-output information, which is classified. However, these omissions are not expected to interfere with the main purpose of this monograph.

The explosion phenomena of the three "low" high-altitude, low-yield events (HA, John, and Yucca) did not differ drastically from that of sea-level or near-sea-level explosions. Compared with the latter, the power-time histories of the fireballs were somewhat shorter, the peak radiances were slightly higher, the time of the minimum (bhangmeter time) was shorter, and the thermal pulse was more intense. While the total prompt-thermal-yield fraction was close to the thermal fraction of similar events near sea level (i.e., 25-30%), the somewhat shorter duration of the pulse is more effective in producing thermal damage. Therefore, the thermal-radiation effects of explosions in the "low" high-altitude domain would be more severe than of sea-level events. In the Yucca event, the early power-time history was optically better resolved than in sea-level events; one could differentiate between the radiative expansion of the fireball, the formation of the hydrodynamic shock, and the debris shock catch-up with the hydrodynamic shock. The fraction of the thermal yield emitted by these three pulses is higher than in the corresponding pulse at sea level, but the relative environmental effects in this yield domain are small. The Tightrope event was also in the low-yield domain. The same comments apply. The thermal pulse was definitely shorter and more intense.

In the next altitude domain, the megaton-range Orange event (Fig. 1) was fired at 43 km. Bluegill (Fig. 2) belongs in the same category, but the yield was lower (i.e., submegaton). In both cases, the phenomenology differed substantially from the lower altitude events. Because of the much lower air density, the x-rays from the source had a larger mean free path (mfp) and the radiative expansion of the fireball was more pronounced. The strong shock still formed early, though it was delayed compared with the strong shock formation in the previously discussed events. The thermal pulse was much shorter, the peak radiances were considerably higher than at sea level, and the apparent bhangmeter minimum appeared as a weak inflection. The time of occurrence does not agree with any popular scaling law. The total prompt thermal-yield fraction was still almost normal. Of great significance, particularly in the case of Orange, was the rise of the debris to altitudes of several hundred kilometers and their subsequent spread and worldwide motion. The effects of this phase of the phenomenology on worldwide communications are reviewed later, as is the motion in the upper stratosphere of ^{102}Rh, a neutron-activated tracer produced in the Orange device.

3

TABLE I

US HIGH-ALTITUDE EVENTS

Event	Place	Date & Time (GCT*)		Local Date and Time	Altitude (feet)	Altitude (km)	Yield
HA	NTS	4/ 6/55	1800	10:00 a.m.	36 620	11.2	3 kt
John	NTS	7/19/57	1400	6:00 a.m.	20 000	6.1	~2 kt
Yucca	Eniwetok	4/28/58	0240	4/28/58 2:40 p.m.[b]	86 000	26.2	Low
Teak	JI	8/ 1/58	1050	7/31/58 11:50 p.m.[c]	252 000	76.8	Megaton Range
Orange	JI	8/12/58	1030	8/11/58 11:30 p.m.[c]	141 000	43	Megaton Range
Argus I	South Atlantic[d]	8/27/58	0230	1:30 a.m.		200 [e]	1-2 kt
Argus II	South Atlantic[d]	8/30/58	0320	2:20 a.m.		240 [e]	1-2 kt
Argus III	South Atlantic[d]	9/ 6/58	2210	9:10 p.m.		540 ± 100[e]	1-2 kt
Starfish	JI	7/ 9/62	0900	7/ 8/62 10:00 p.m.[c]		400	1.4 Mt
Checkmate	JI	10/20/62	0830	10/19/62 9:30 p.m.[c]		Tens	Low
Bluegill	JI	10/26/62	1000	10/25/62 11:00 p.m.[c]		Tens	Submegaton
Kingfish	JI	11/ 1/62	1210	11/ 1/62 1:10 a.m.[c]		Tens	Submegaton
Tightrope	JI	11/ 4/62	0730	11/ 3/62 8:30 p.m.[c]		Tens	Low

[a] GCT = Greenwich Central Time (= U. T.)

[b] Eniwetok Daylight Saving Time

[c] JI Time = Hawaii Time Minus One Hour

[d] Argus I   12°W 38°S
   Argus II   8°W 50°S
   Argus III 10°W 50°S

[e] References for Argus revisions:

1. N. C. Christofolis, "The Argus Experiment," J. Geophys. Res. 64, 869 (1959).
2. W. N. Hess, *The Radiation Belt and Magnetosphere* (Blaisdell Publishing Co., Waltham, MA 1968).
3. R. W. Kilb, "Analysis of Argus III Photographic Data (U)," Unclassified Parts of Mission Research Corporation report MRC-R-112 (January 1974).
4. R. W. Kilb, "Analysis of Argus II All-sky Photographs (U)," Unclassified Parts of Mission Research Corporation report MRC-R-176 (March 1975).

In the megaton-range Teak event (Fig. 3), fired at 76.8 km, radiative expansion was the dominant feature of the early phase. The so-called x-ray fireball had radial dimensions of the order of 10 km. Almost all the prompt thermal radiation was emitted during this period. The total thermal-yield fraction was only slightly lower than for a similar sea-level event, but the pulse was much sharper and the peak radiance very much higher. The shock formed late (order of one second) and the shock phenomena—air shock and debris shock—were visible to radial distances in excess of 500 km. The debris were seen to rise much faster, understandably, than in the case of Orange. Consequently, worldwide communication deterioration began much earlier. The fission-product beta rays formed well-defined, field-aligned auroras going north and south.

Kingfish (Fig. 4), a submegaton explosion fired at a higher altitude than Teak but still below the horizon as seen from Mt. Haleakala on Maui Island, had many similarities with Teak. Because of the still-thinner air, the effects of the magnetic pressure were pronounced at late times, leading to striated field alignment of the debris, besides the early formation of beta-ray excited, bright auroral pencils. The thermal fraction was lower because the surrounding air was heated to lower temperatures at which air is a poor radiator. Communication interference was not very severe.

Checkmate (Fig. 5) was a low-yield explosion at a still higher altitude. The burst point was just about visible from the Hawaiian Islands. The effects of the magnetic field on debris motion were even more pronounced than in the case of Kingfish. The prompt thermal output was low. Effects on communications were mainly local.

Going up in altitude, one must mention the three Argus experiments with yields of 1-2 kt fired from shipboard in the South Atlantic to altitudes of ~200, 240, and 540 km. Planned and executed by the Department of Defense, the operation was originally classified. However, the observations were of great scientific value, and a year later the experiments and data were declassified and reported in the open literature. The events produced the first artificial radiation belts, shortly after the discovery by Van Allen of the natural belts.

Finally, Starfish was fired with a yield of 1.4 Mt at 400 km altitude above Johnston Island (JI) (next to Argus III the highest event). While many of the results were of military value, Starfish was also an experiment of worldwide scientific interest. Yield, altitude, and time of event were announced prior to the event.

At Starfish altitude, magnetic pressure and air-particle pressure are of about the same magnitude; therefore, the field effects play a very strong role from the earliest time on in the event. Indeed, the debris motion was largely governed by the magnetic field (Fig. 6). A large fraction of the debris moved swiftly down the fieldlines, to be stopped at ~100 km altitude in the northern conjugate area. Another fraction moved to the southern conjugate area. Other debris at first remained near the burst area, spreading over distances of the order of 500 km. Finally, a small percentage jetted to altitudes of 1 000-2 000 km or more, leading to densely populated long-lived artificial radiation belts (Fig. 7). The belts interfered with some then-current observations of natural magnetospheric and astrophysical phenomena, but their study contributed greatly to our understanding of many physical processes occurring in this space. The prompt thermal output of the Starfish event was very small—in fact, insignificant. Radio communication interference was less severe than after Teak and Orange, owing not only to the lower yield, but more to differences in phenomenology.

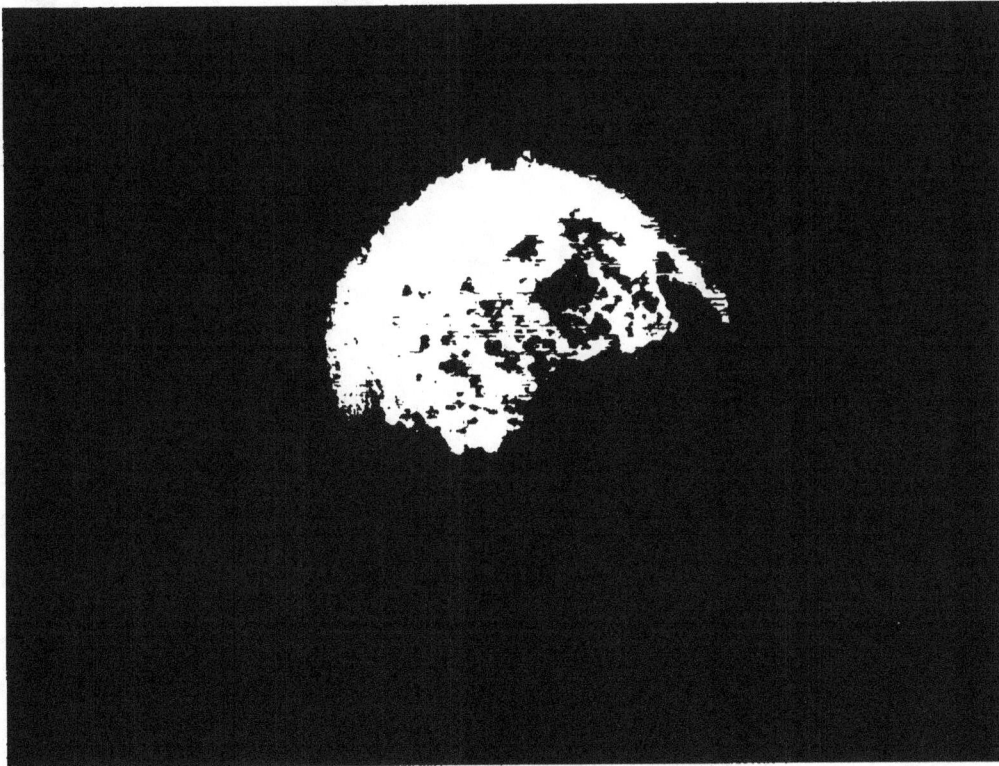

Fig. 2.
Bluegill Event seen from high-flying aircraft.

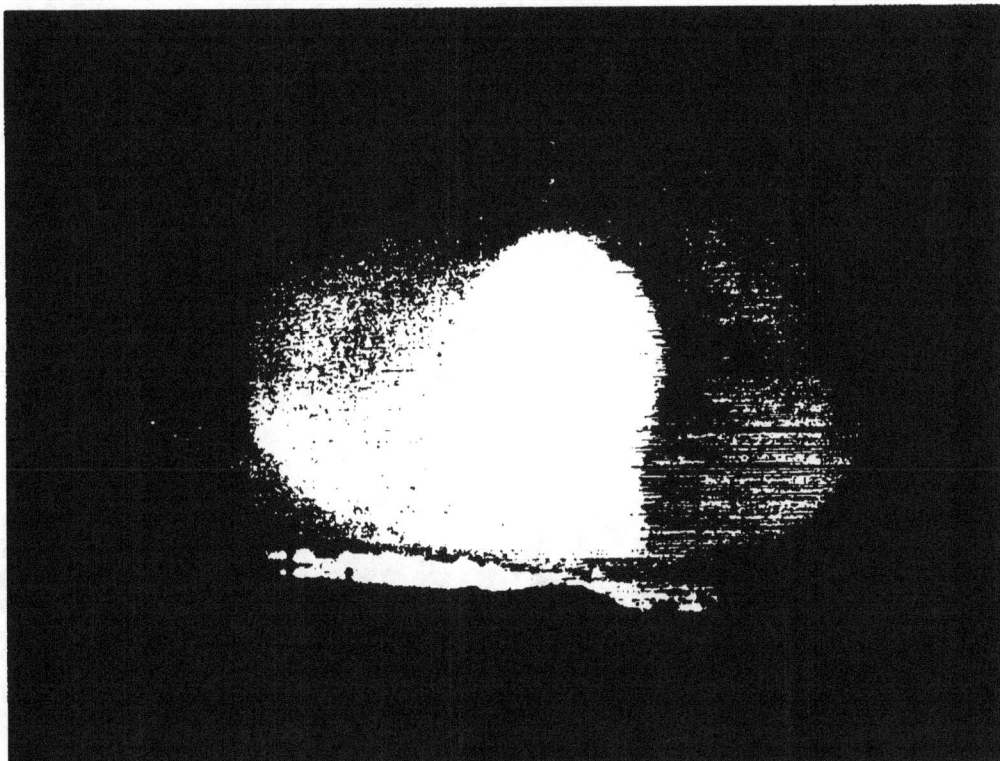

Fig. 3.
Teak Event seen from top of Mount Haleakala (Maui) at approximately H + 1 minute.

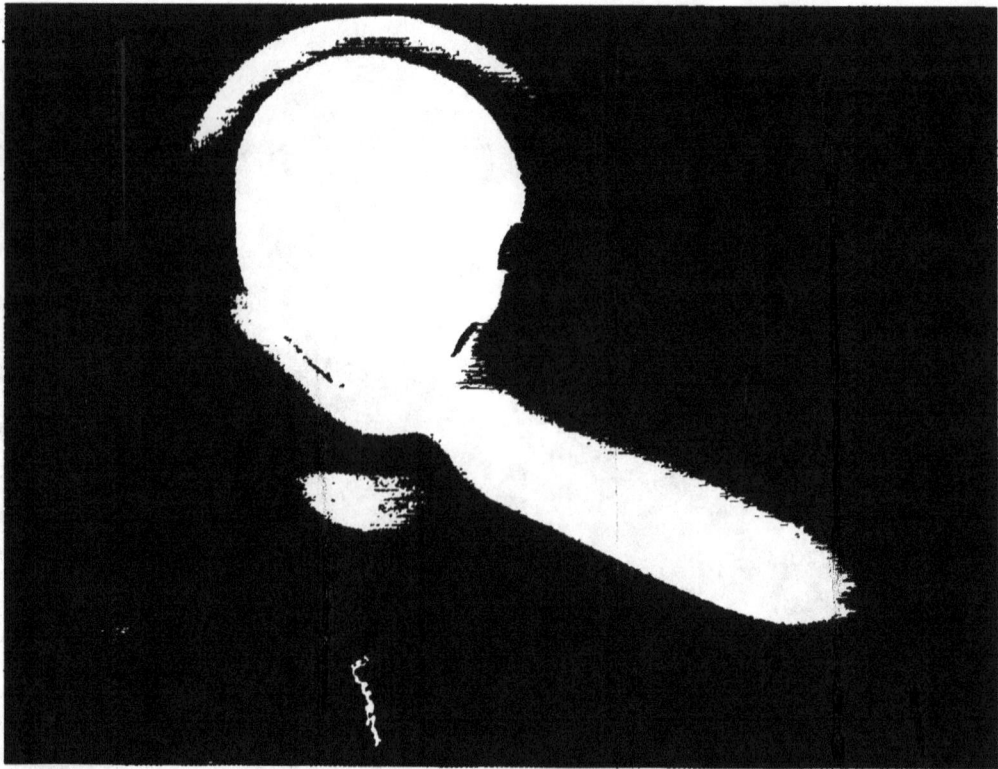

Fig. 4.
Kingfish Event seen from high-flying aircraft.

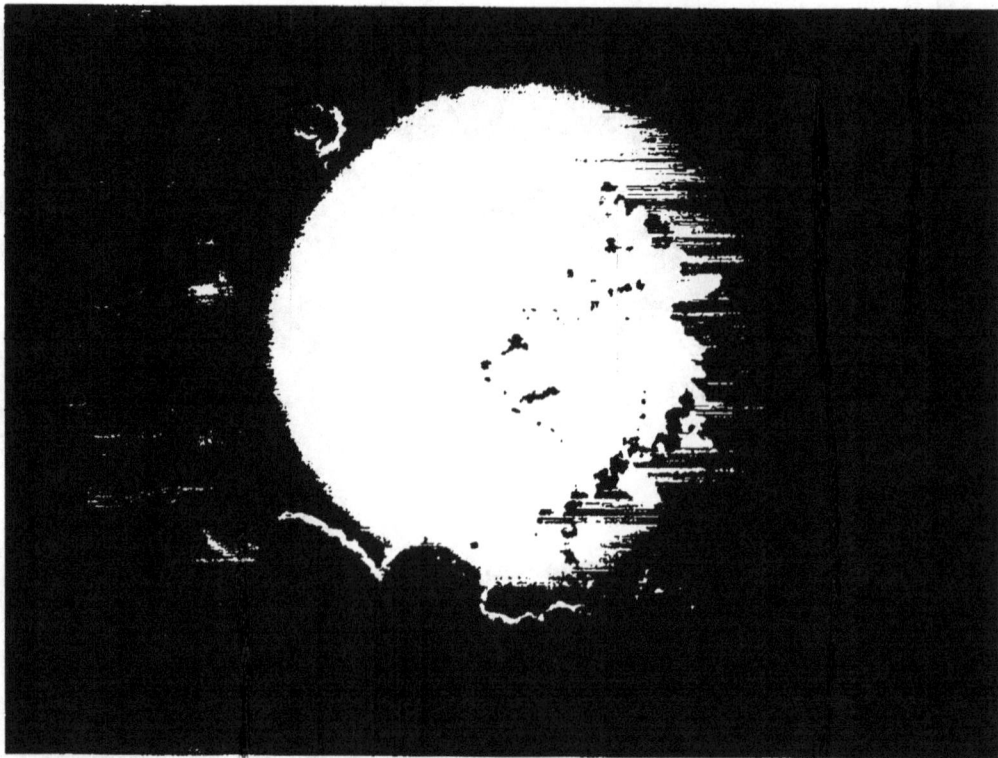

Fig. 5.
Checkmate Event seen from Johnston Island.

7

Fig. 6a.
Starfish event. Air fluorescence excited by magnetic-field-aligned debris particles seen from aircraft at approximately H + 3 minutes.

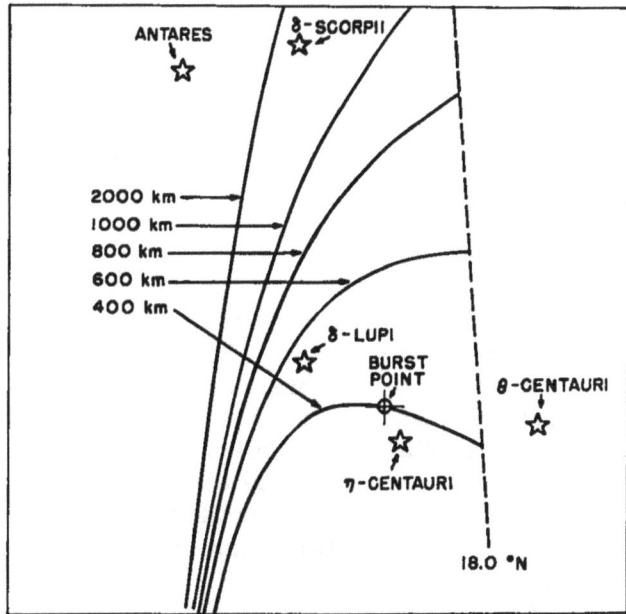

Fig. 6b.
Projection of magnetic field lines into field of view of camera. Compare with Fig. 6a.

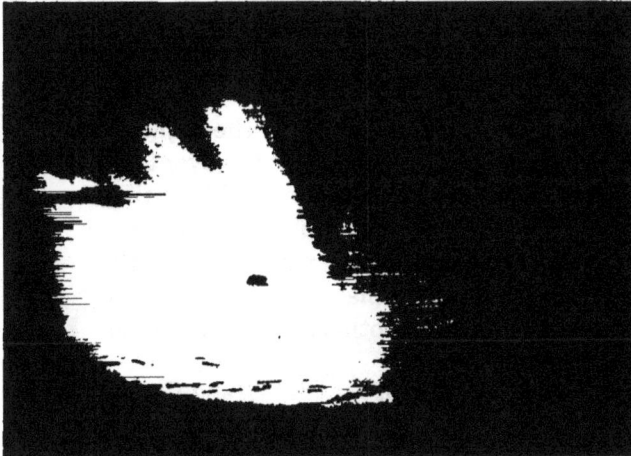

Fig. 7a.
Starfish Event seen from Christmas Island. Air fluorescence excited by debris motion at approximately H + 1 minute.

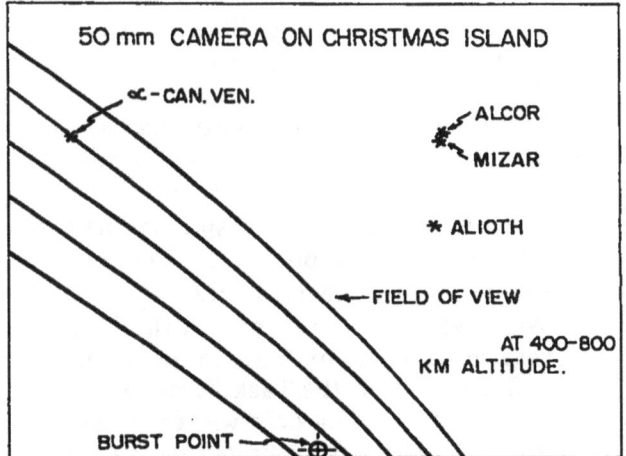

Fig. 7b.
Projection of undisturbed magnetic field lines into field of view of camera. Compare with Fig. 7a and note crossing of field lines by debris.

CHAPTER II

PROMPT THERMAL RADIATION.

Prompt thermal radiation is defined as that part of the fireball radiation which is emitted in times of the order of a few seconds or less. This radiation, $\lambda > 3\,200$ Å, is transmitted by undisturbed air to long distances. If intense enough, it can produce damage to materials and serious injury to humans. The power peak of the radiation[1] at sea level occurs roughly at the time of the second maximum: $t_{2nd\ max} \approx 0.03 \sqrt{Y}$ seconds, where Y is in kilotons. At 10 times second maximum time, the rate of emission has decreased by one or several powers of 10. In the case of a 1-Mt explosion at sea level, the second maximum occurs at about one second; and by 10 seconds, 25-30% of the total yield has radiated away from the fireball.

Glasstone[1] gives data for the approximate radiant exposure for ignition of materials and dry forest fuels for pulse durations of approximately one second and longer. For instance, the ignition exposure for shredded newspapers is 4 cal/cm² for a two-second pulse (10 x second-maximum time) from a 40-kt explosion at or near sea level. For shorter pulses, the ignition threshold is lower.

At higher altitudes, there is a change in the physical processes that control the interaction between the nuclear radiation and the surrounding air. As described briefly in Chapter I, the radiative expansion becomes more dominant and the thermal pulse becomes sharper. At Teak altitude, for instance, the significant time scale is now a few tens of milliseconds for megaton-size yields. Up to this altitude, the fraction of the total yield that appears in the thermal pulse has changed very little. However, because of the sharpness of the pulse, the damage produced by a given caloric impulse is more severe.

At still higher altitudes, from about 100 km on, the prompt thermal output becomes significantly lower. It is of the order of 10% of the yield at 100 km, and drops steeply to small fractions of a per cent at Starfish altitude. Thus, the effects of the gross thermal radiation on the natural environment were insignificant for the majority of the US high-altitude explosions. There were two exceptions.

In the case of Teak, we expected a maximum dose of 1 cal/cm² on JI and on the adjacent bird refuge on Sand Island. No thermal damage was expected. However, after the event, we observed quite a few birds sitting or hopping on JI docks in a helpless manner. Either they had been blinded or they were unable to dive for fish, their major food supply, because the ethereal oils which protect their feathers from getting water-soaked had been boiled off by the thermal pulse. Otherwise, the only thermal effect I am aware of was that my colleague, Don Westervelt, who had watched the burst with dark goggles but was otherwise unprotected, received a slight sunburn on his forehead and his forearms.

Subsequently, the Task Force took measures to protect the birds from the "Orange" thermal radiation. The dose was expected to be two to three times that of Teak. An artificial smoke screen was generated to cover Sand Island at explosion time. But Orange was fired above a rather dense cloud cover—perhaps fortunate for the birds escaping from the smoke, but unfortunate for groundbased diagnostic optical observations. The exact caloric doses measured on the ground on all US atmospheric tests are still classified. Approximate numbers are given in Table II.

(It might be worth mentioning that very high yields[2] [50-100 Mt] exploded in the Orange-Bluegill-Teak altitude domain would be most effective in starting fires over large areas.)

A brief historical note might be of some interest: The thermal-damage problem was considered as early as May 1957. At the Eniwetok Planning Board meeting on May 14, 1957, Ogle

9

TABLE II

PROMPT THERMAL DOSE FROM HIGH-ALTITUDE EXPLOSIONS

Event[a]	Altitude (km)	Approximate[b] Thermal Dose at Ground Zero (cal/cm^2)	Approximate[b] Duration of Main Pulse (ms)
HA	11.2	4×10^{-2}	300
John	6.1	6×10^{-2}	300
Yucca	26.2	$< 10^{-2}$	150
Teak	76.8	1.0	100
Orange	43	3.0	150
Starfish	400	$< 10^{-5}$	< 1
Checkmate	Tens	$< 10^{-6}$	< 1
Bluegill	Tens	10^{-1}	100
Kingfish	Tens	2×10^{-2}	150
Tightrope	Tens	$< 10^{-1}$	150

[a]Yields are listed in Table I.
[b]All numbers are approximate numbers. They are adequate for environmental-effects purposes. Only for the Teak and Orange events is the prompt thermal dose of significant magnitude.

reported that there was no danger of thermal damage, at least for Rongerik/Rongelap. One June 26, 1957, Duane C. Sewell wrote to A. C. Graves in regard to the proposed Teak event, "...we have held several discussions here among UCRL (Livermore) personnel...and also Mike May and Tom Wainwright have discussed this problem with Al Latter and some of his people at Rand. This group has been unable to devise a model which we agree could be relied upon with any degree of confidence to predict the thermal flux on the ground directly under this shot."

10

CHAPTER III

FLASH BLINDNESS AND EYEBURN.
EYEBURN CASE HISTORIES.
EYEBURN HAZARD AND OPERATIONAL PROBLEMS.

Flash Blindness

Flash blindness is defined as the temporary loss of vision resulting from photostress. Photostress results from exposure to a high-intensity light source from which an after-image develops.[3] An extensive literature, including results from several laboratory tests, describes a large variety of exposure conditions.[4,5]

The recovery time depends on numerous variables such as flash luminance, flash duration, source spectrum and geometry, flash distance, and degree of eye-adaptation of the observer. The regeneration of the visual pigments and an apparent automatic brightness control which reduces the sensitivity of the bleached retinal area are age-dependent; there are also wide variations in responses from subject to subject.

I am not aware of flash-blindness studies conducted during high-altitude explosions. However, several measurements were taken during other nuclear test operations.[3,4] They apply generally, although the conditions of the observers at the times of the test are not clearly described. Probably useful numbers are as follows. For an incident energy of ~ 0.01 cal/cm^2 at the cornea, the recovery to 0.1-0.3 visual acuity took 72 and 90 seconds for one subject; times to read aircraft instruments with standard edge lighting and red floodlighting were 10-12 seconds for two other subjects. Two subjects were behind sandblasted aircraft windows; they required 90 seconds to recover to 0.1 visual acuity. Scaling to the Teak event, similar condition would exist at slant distances of the order of 400 km.* Thus, for events like Teak and Orange, disturbing effects will occur at night whenever the observer faces the burst and when the fireball is above the horizon.

Long recovery times are, of course, a threat to commercial and, even more so, to military tactical air operations. Consequently, photochemical shutters which close within one millisecond or less can reduce the duration of the "flash-out" to one second or less.

Flash blindness does not involve the focusing of the source on the retina. If focusing occurs, permanent damage (eyeburn) may result. Circumstances producing eyeburn also cause flash blindness, but flash blindness does not necessarily involve eyeburn.

Eyeburn (Chorioretinal Burns)

Retinal eyeburn is the result of thermal-energy deposition in the *image* of an intense source of light—in contrast to flash blindness, which is the temporary incapacitation of vision by an unfocused flash. If the dose is above a certain safe limit, the damage to the retina is irreversible. Permanent retinal lesions cause scotomas, or blind spots.

On Teak, chorioretinal burns occurred on rabbits at distances exceeding 480 km. The rabbit's eyes were focused on the fireball. Note also that the rabbit's blink period is 300 milliseconds versus 150 milliseconds for the average human.

11

The existence of a nuclear-burst eyeburn problem was recognized before the Trinity event. It is well known that solar-eclipse observers have suffered mostly minor damage due to carelessness. Predictions of the early radiance of the fireball yielded numbers much in excess of the radiance of the solar photosphere. Consequently, observers used high-density dark goggles for protection.

Use of goggles with density 4, attenuating the light fluxes by a factor of 10 000, became the rule in subsequent tests. In the 1950's, semiquantitative studies were made during several US test operations. More quantitative investigations were made in several laboratories, such as at the Medical college of Virginia (Dr. W. T. Ham and associates) and the Ophthalmology Department of the United States Air Force School of Aerospace Medicine, Brooks Air Force Base, Texas, and others. The burn threshold depends on many factors, such as image diameter, rate of thermal energy deposition, total dose, and spectral characteristics of the source. One must differentiate between a *threshold* dose—the dose that produces a detectable burn in 50% of the cases—and the *safe* dose. For instance, for diameters on the retina of a few hundred micrometers and exposure times of ~100 microseconds, Ham and associates[6] in 1963 found a rabbit threshold of 0.2 cal/cm^2, Miller and White's[7] threshold on primates was 0.1 cal/cm^2, and Allen and associates' (Ref. 5) threshold on rabbit eyes was reported to be 0.1 cal/cm^2. Ham's[6] safe value was 0.05 cal/cm^2.

During the planning stages of Dominic in late 1961 and early 1962, Ham, Ogle, Shlaer, and Hoerlin conferred repeatedly and accepted a threshold tolerance of 0.05 cal/cm^2 for small image sizes (approximately 50 μm) and exposure durations of microseconds. This tolerance was based on Dr. W. T. Ham's suggestion at that time that a prompt temperature increase of 10 to 20°C in the affected area—i.e., in the pigmented epithelium—was the tolerance level.*

For theoretical treatments, considering experimental data mainly on rabbits' eyes, reference is made to work by Hoerlin, Skumanich, and Westervelt,[8] by Mayer and Ritchey,[9] by R. Cowan,[10] and the latest, and I believe the most advanced, study by Zinn, Hyer, and Forest.[11] In the course of these studies, the safe dosage levels were more clearly established. A temperature rise of not more than 5°C in the 10-μm-thick pigmented epithelium of the retina is now considered safe. If the temperature rise is 20°C or more, a burn results. This criterion was then applied to calculate safe dosage levels in terms of cal/cm^2 incident on the retina for different image diameters, exposure times, and source temperatures. Such data are published:[11] for instance, 1 cal/cm^2 incident over a period of 100 milliseconds and an image size of 100 micrometers would be safe; however, if a dose is delivered in 10 milliseconds, only about 0.2 cal/cm^2 would be safe. These numbers apply to source temperatures in the 5 000 to 15 000 K range.

Most of these data apply to sea-level or near-sea-level explosions, when the important radiating temperatures are in this particular regime. At higher altitudes, the physics of light emission is different. There was already an indication of such a difference at the time of the HA shot during Operation Teapot. All visual observers (with goggles) agreed that the fireball appeared more intensely bright than in events of similar yield fired at lower altitude.

As discussed briefly in Chapter II of this report, the optical power versus time history of high-altitude explosions changes with altitude. Generally, with increasing altitude, the thermal-pulse duration decreases; i.e., the flux rate increases, and thermal conduction in and near the retinal image is consequently less effective in reducing the temperature increase. This effect is of moderate significance for the low-yield, moderate-altitude events HA, John, and Yucca. It is very significant for the 50- to 150-km altitude domain of Orange, Bluegill, Teak, Kingfish, and Checkmate. In the latter cases, the thermal-pulse durations are of the same order of magnitude or shorter than the natural blink period which, for the average person, is about 150 milliseconds. Furthermore, the atmospheric attenuation is normally

---

The pre-Dominic notes report about "threshold tolerance" and "tolerance level." What was probably meant is "threshold dose."

12

much less for a given distance than in the case of sea-level or near-sea-level explosions. Consequently, the eye-damage hazard is more severe. Fortunately, the seriousness of this problem was recognized during the early planning stages of the Teak and Orange events. It seems worthwhile to document that phase of the eyeburn hazard and its operational consequences separately in one of the following sections.

At still higher altitudes, as with the Starfish event (400 km above JI), the fireball phenomenology is changed again: the main thermal x-ray energy radiated by the source has a very long mfp and is absorbed over a very large volume of air at about 100 km altitude, thus producing mainly air fluorescence, little heating of the air, and relatively low radiance—i.e., low optical power per unit area, with no hazard from this source. The fraction of the total energy release that resides in the internal energy of the expanding bomb debris, part of which is emitted in a sharp pulse, is not insignificant. However, the source is small, and at distances of 400 km and more it is not resolved by the eye. Thus, the burn hazard is lower at the greater altitudes. The prediction of the thermal output from this source as a function of time and diameter was done by Longmire. The pre-event concern about possible effects on observers in the Hawaiian Islands* was resolved after careful review of the problem. There could be no above-threshold exposure in the Islands, and indeed no eye damage nor other physiological inconveniences were reported. The burst was observed from several beaches and mountain tops at slant distances in excess of 1 000 km.

The approximate retinal dose an observer could have received at ground zero, the safe dose, and the approximate safe slant distance are shown in Table III for all major high-altitude explosions. It is evident that on clear nights or days, only Starfish and Checkmate could have been viewed safely from directly underneath or from JI.

Eyeburn Case Histories

A total of nine case histories of eyeburn produced by US nuclear test explosions have been reported.[4] Two of these occurred accidentally on JI during the Bluegill event, which was fired at night. Reference 4 describes these cases as follows:

"...The burns were sustained at a slant range of about thirty miles. Neither individual had his protective goggles on during the detonation. The pulse characteristics of this particular detonation...had trailed off to low levels well before a blink reflex could occur. Peak irradiance at the ground station was between two and three watts/cm².** This means that the blink reflex would have been of no protective value and that the injured individuals had to be fixating at the exact detonation point when the detonation occurred. One case does give evidence which suggests that the eyeball may have been in motion during the damaging phase of the fireball, since a small tail-like extension was observed on the lesion. However, there is also a remote possibility that the two burn victims could have been burned by a specular reflection rather than the direct image. Such reflections could occur from a wristwatch face or any of a variety of shiny metal or glass surfaces.†

"The clinical data for these latter two burn cases is fairly typical, except that the damage to central vision was more pronounced than the six low-altitude cases

---

The countdown was monitored and rebroadcast by commercial radio stations in Honolulu.

**According to my calculations, the dose could have been as high as 8.5 cal/cm² (no reflection); the safe dose was 0.2 cal/cm² (see Table III).*

†*Reflection from a water puddle is another possibility. Rain had fallen before the event.*

13

TABLE III

RETINAL DOSE AT GROUND ZERO, SAFE DOSE, AND APPROXIMATE SAFE SLANT DISTANCE

| | Ground Zero | | Approximate | |
Event[a]	Estimated Dose[b] (cal/cm²)	Safe Dose (cal/cm²)	Safe Slant Distance (km)	Eye Adaptation
HA	3.2	0.5	30	Daylight
John	4	0.9	25	Daylight
Yucca	1.7	0.6	100	Daylight
Teak	23	0.2	720	Night
Orange	50	0.4	2° Elevation[d]	Night
Starfish		Safe any distance larger than 400		Night
Checkmate		Safe any distance larger than 200		Night
Bluegill	8.5[c]	0.2	2° Elevation[d]	Night
Kingfish	0.1	0.04	200	Night
Tightrope	20	0.3	250	Night

[a]For burst altitude and approximate pulse duration, see Tables I and II.
[b]For observer at ground zero.
[c]On Johnston Island, slant range ~60 km.
[d]From sea level.

cited previously. In the first case, acuity for central vision was 20/400 initially, but returned to 20/25 by six months. The second victim was less fortunate, as central vision did not improve beyond 20/60. The lesion diameters were 0.35 and 0.50 mm* respectively. Both individuals noted immediate visual disturbances, but neither was incapacitated. In a recent review[12] of these two cases, the fact that chorioretinal burns on or near the fovea do not necessarily cause complete blindness was emphasized. Both *size* and *location* of the lesion determine visual impairment."

The recent review[12] referred to above describes in considerable detail the observation of the two patients by ophthalmologists during more than 6 months, first at Tripler General Hospital and later at the USAF School of Aerospace Medicine. After 6 months, the Air Force sergeant (Case No. 1) performed unusually well in his job and had minimal subjective complaints. "His reading ability was good; when he held his eyes stationary, he was aware of a very small central negative scotoma which blanked out individual letters." "The U.S. Navy petty officer (Case No. 2) was not as effective in his assigned duties, and his visual findings were somewhat more severe than those of the Air Force sergeant." In this case "...the fovea was destroyed, and there is no doubt that more energy was absorbed." "... He has been discharged by the Navy with a disability ratable at 30%; however, there are many gainful occupations that he can perform very capably."

*According to my calculation, the diameter of the "hot" image on the retina was 0.35 mm for a focal length of 15 mm.

Teak Eyeburn Hazard and Operational Problems (A Historical Review)

According to available records, the Teak eyeburn problem was first discussed on May 11, 1957, during a meeting at the Control Point, Mercury, Nevada Test Site (NTS), with W. Ogle, H. Stewart, and H. Hoerlin participating (entry in Hoerlin's notebook dated May 11, 1957). The eyeburn hazard was considered to be serious. The consensus was that further study was needed. It was felt that dark goggles were certainly needed at Rongerik, an inhabited atoll 250 km from Bikini. The seriousness of the problem was subsequently relayed to the Eniwetok Planning Board meeting at NTS on May 14, 1957, by Ogle. On June 29, 1957, H. E. Parsons (a Department of Defense [DOD] representative at the Eniwetok Planning Board) wrote to A. Graves of the Los Alamos Scientific Laboratory (LASL): "An evaluation of the flash blindness hazard has not been attempted for lack of understanding of the phenomena affecting brightness...."

In July and August, H. A. Bethe delivered a series of some 10 lectures at LASL to J-Division staff members on the physics of megaton-range high-altitude events.[13] On the basis of these discussions, theoretical calculations of the Teak phenomenology were started in Group J-10, mainly by A. Skumanich and F. Jahoda. First data were obtained November 1957. Preliminary calculations of anticipated fluxes on the retina of the dark-adapted eye directly under Teak predicted a dose of 27 cal/cm^2 during the blink period of 150 milliseconds, with a possibility of a hotter spot in the center. The damage threshold was then believed to be 3 cal/cm^2 (entry in Hoerlin's notebook dated January 17, 1958). More advanced theoretical numbers were reported by Skumanich on March 3, 1958.[14] On March 13, 1958, W. Ogle sent a telegraphic message to General A. Luedecke, the Task Force Commander, pointing out the seriousness of actual eyeburn danger: the possible danger radius at sea level was quoted to be 540 miles on exceptionally clear days. This message was supplemented shortly afterward by two Hardtack eyeburn documents written by W. Ogle.[15] The main concern of the Task Force was protection of the Marshall Island natives. Approximately 11 000 Polynesians lived within 400 nautical miles of Bikini Atoll, where the event was planned to take place. Their main livelihood is fishing, frequently at night. The probability of their observing the rocket launch, following the track, and then being focused at or near the burst point was considered high. It was also felt that it would be impossible to keep all the natives, dispersed over 20 inhabited islands, under control and/or equip them with goggles. Consequently, on April 9, 1958, the decision was reached to move the Teak and Orange events to the practically unpopulated JI area. (It is not clear to me at this time who was involved in the decision. The high-altitude events were proposed by the Department of Defense [DOD], more specifically by the Air Force. The Nuclear Panel of the Scientific Advisory Board of the Air Force was probably the driving force. The Joint Chiefs of Staff had to approve the general schedule. The Air Force Special Weapons Project [AFSWP], located at Kirtland Air Force Base, was the DOD's executing agency. The Atomic Energy Commission [AEC] was involved, but so far as I know mainly as a needed participant. In any case, the move to JI was requested by W. Ogle, the Scientific Deputy of the Task Force; and the decision to execute the move was made by the Commander of the Joint Task Force.)

As to more technical details, in arriving at a danger radius of 540 miles, we considered a variety of assumptions, such as daylight versus nighttime firing, and "high" and "probable" atmospheric transmission. Some burn thresholds (cal/cm^2) for rabbit eyes were known, from Dr. W. T. Ham's work,[6] as functions of image diameters and exposure times. Several source diameters, thermal yields, and durations of the thermal pulse were assumed. In the end, the conservative approach was taken—namely, a source diameter of 4 km and a pulse duration of 30 milliseconds; 3 cal/cm^2 were taken as the "maximum allowable dose" for the retinal image size at 540 miles horizontal distance from ground zero for nighttime conditions and an exceptionally clear atmosphere.

15

As discussed in a preceding section, in the years following Hardtack the problem was treated in greater depth. This work, much of which was done on primates (rhesus monkeys), led to a reduction of the threshold and safe dose exposure. We believe now that irreversible damage occurs for a temperature increase of 20°C, while a 5°C temperature rise is safe. Applying these criteria to the Teak case, the threshold (20°C) dose at ground zero would reduce from 3 cal/cm² to 1 cal/cm² on the retina and the safe dose to 0.2 cal/cm². Then, taking the *post*event source data and assuming an exceptionally clear day, the safe slant distance would have been 450 statute miles.

CHAPTER IV

EFFECTS ON RADIO COMMUNICATIONS.

General

Ionization produced by the high-altitude events caused degradation of radio communications over large areas of the Pacific. The most severe effects occurred after the Teak and Orange events; they were less severe after Starfish and relatively moderate as a consequence of Checkmate, Kingfish, Bluegill, and Tightrope.

The ionization was caused mainly, but not exclusively, by fission-product gamma rays, and, in more local areas, by beta radiation. The extent and intensity of ionization was governed by the location of the debris and by the fission yield. When the debris rise from the burst location to higher altitudes, they spread in space, and the gamma rays, because of their long mfp, cover increasingly larger geographic areas. The debris cloud as such is also highly ionized because of the short range of the beta particles, although many betas escape and produce ionization in conjugate areas.

The Teak and Orange events had the highest yields. The Teak debris rose relatively fast, reaching altitudes of 500 km in about 20 minutes. Little direct quantitative information about the subsequent motion exists, but both the actual debris cloud and the associated gamma-ray effects were sources of serious communication blackouts in the South Pacific, New Zealand, and Australia, mainly in the MF* and HF* ranges. Some details are described in later sections.

The Orange debris rose more slowly from its lower burst altitude. Therefore, it took longer to affect the D-layer horizon. The onset of severe degradation was delayed, but after it occurred it was as strong as and generally of longer duration than after Teak. The main body of these debris rose to 150-250 km altitude; however, there are indications that fractions rose higher, perhaps to 500 km.

The communication interference patterns after Starfish were different from those encountered after Teak and Orange. There was little delay in the onset of the initial absorption. This difference is caused by the differences in burst altitude. While the x-rays emitted by the Orange and Teak devices deposited their energies in the air close-in, the Starfish x-rays traveled long distances. The effects of their prompt energy deposition in the upper D-layer and of gammas at 25-30 km were not very long-lasting. We assume that about 30% of the debris were then spread over diameters of 1 000 km or more near the burst altitude, acting as fission-product gamma-ray sources. Another 30% each of the debris moved along the magnetic field lines to the northern and southern conjugate areas where large debris patches

---

Nomenclature used by communication engineers:

VLF - Very Low Frequency	*<30 kHz*
LF - Low Frequency	*~30 kHz to 300 kHz*
MF - Medium Frequency	*~300 kHz to 3 MHz*
HF - High Frequency	*~3 MHz to 30 MHz*
VHF - Very High Frequency	*~30 MHz to 300 MHz*

(~500 x 1 000 km initially) were formed, producing near this space large volumes of ionization by fission beta particles; the gamma rays produced lower but still significant ion densities over still larger volumes. Later, the debris patches were observed to rise from the conjugate areas, and to spread over still larger areas of the Pacific. Thus, the Starfish debris space-time history differed significantly from the Teak and Orange histories.

While D-layer absorption of radio frequencies was the main cause of the communication blackout, it is interesting to note that long-distance VHF transmissions improved after all three events in several areas, particularly at night, because of the increased electron densities in the E- and F-regions.

The following sections provide details of the interferences. The literature for the Teak, Orange, and Starfish events is so extensive that only a selection of the most severe and interesting occurrences is made here. Very little has been published in the open literature on communication problems after the other events. The effects were of a more local nature, and the fact that precise yields and altitudes remain classified did not facilitate the interpretation of whatever was observed. A brief description of what has transpired will, however, be given.

Teak and Orange Effects

Johnston Island. After the Teak burst, the island communications were cut off for many hours; unfortunately, I have been unable so far to find detailed records. However, I was present on the island and remember not so much the difficulties encountered by the JI communication people in making contact with the outside world but rather the desperate attempts of other transmitting stations to obtain a response from JI. One of the first transmissions actually received at JI in the morning hours after the event was "Are you still there?"

Honolulu had serious difficulties in maintaining air travel services. Indeed, they had to be suspended for many hours because of the failure of long-wave communications. H. P. Williams[16] provided the following summary:

Hawaii.

LF and MF Propagation. A serious interruption of LF and MF transmission occurred. Below 1 MHz, the nighttime absorption continued for three days. Above this frequency, the absorption had decreased by the next night. This applies to Teak. In the case of Orange, the persistence was reduced to one day.

HF Propagation. A complete blackout started at about 20 minutes after the explosion in the case of Teak and at plus five hours after Orange.

VHF Propagation. There seems to be no mention in the unclassified literature of the effects of Teak and Orange on VHF ionoscatter propagation. The possible effects are discussed by Williams[16] in the light of known solar-flare events. He concludes that after shots of the Teak and Orange type, absorption of VHF frequencies in the D-region increases. While scattering at altitudes of ~90 km increases, cosmic noise decreases; thus, the signal-to-noise ratio improves. It is concluded, therefore, that VHF links using ionoscatter or meteor-scatter propagation would have escaped the severe blackouts experienced with MF and HF transmission.

Ionosonde Data. The National Bureau of Standards (NBS)[17] operated an ionosonde at Maui, Hawaii; vertical-incidence ionograms were obtained routinely every 15 minutes in the frequency ranges from 1 to 25 MHz. After Teak, "complete blackout"—ie., no reflected signals above 1 MHz—occurred from H + 25 minutes to H + 3 hours and again at H + 4 hours for 15

18

BODY-24

minutes. After Orange, "total blackout" occurred at H + 5 hours and 15 minutes, lasting for 2 hours; and partial to complete blackout lasted for another 2 hours and 45 minutes.

South Pacific Data. G. C. Andrew[18] reports as follows:

Teak.

MF - HF: At Rarotonga* nighttime reception of MF broadcast stations was impossible for some five days after Teak. There was a complete blackout of all communication frequencies in use for commercial aircraft and broadcasting services.

VHF: However, "phenomenally good high frequency communication" became possible because of the abnormally high ionization density in the F-layer. Signals on 30 MHz and above were heard over long distances even at night.

LF: Low-frequency radio signals were also heard during daylight hours over long distances.

Orange.

MF - HF: After Orange, absorption of MF broadcast-station signals was even greater than that following Teak. A complete blackout of these stations lasted for a week. The fade-out of HF radio signals in the Pacific lasted, however, for a shorter period. In Australia, periods of severe attenuation of MF sky-wave signals occurred, lasting one or two days and extending from the 2nd to the 10th day after each explosion.

VHF: On the other hand, the first two-way contacts ever established on 50 MHz between Rarotonga and Hawaii, a distance of about 5 000 km, were made over a path of complete darkness, "presumably because the atomic explosion produced clouds of high ionization that extended or drifted over an area sufficiently large to permit multiple-hop propagation."[19]

Near Wellington, New Zealand, the BBC transmissions from England and the relay stations in Singapore at frequencies between ~21 and 26 MHz were enhanced at various periods after the Teak and Orange events.

Communication Links Across the Pacific. Obayashi, Coroniti, and Pierce[19] published changes in signal strength over the HF links from Japan to Honolulu and to San Francisco. Williams[16] summarizes the main features as follows:

	Japan-Honolulu (10 MHz)	Japan-San Francisco (14 MHz)
Teak	40 db drop for 6 hours	40 db drop ±20 db
Orange	20 db drop after 5 hours	±10 db variation

Rarotonga is at the southern end of the Cook Island group, at ~20°S, almost straight south of the Hawaiian chain.

There exists additional specific information in the literature. The important fact is that not only channels passing in the vicinity of JI but also channels at very large distances from the burst area were affected, indicating strong disturbances from the D-layer up to the F-layer. For instance, the San Francisco-Japan transmission link passes 3 600 km away from JI.

Starfish Effects

Johnston Island, Northern Hemisphere. In the HF range, total blackout occurred on JI only for a short time; moderate interference lasted for several hours. Absorption increased at sunrise.

The debris patch in the northern conjugate area should have affected communications in the French Frigate Shoal, Midway, and Wake areas; also on board ship—the "DAMP" ship, for instance. The information reposes in the classified literature.

The signal strength of Radio Australia in Melbourne on 11.7 MHz was measured at Lexington, Massachusetts.[20] The short path crosses within 2 300 km of JI, but passes through the southern conjugate area. First, there was a sharp drop in signal strength by 10 db, lasting two minutes; then came a 20-db drop (total) for five additional minutes. Recovery after plus seven minutes was almost complete.

Australia, New Zealand, Cook Island Areas. Apparently, the strongest communication degradation occurred in this area caused by the motion of a large debris fraction into the conjugate area[21] and its subsequent expansion and drift. The following information has been extracted from the literature.[18]

Wellington, New Zealand, monitored countdown from JI at 12.020 MHz. After explosion time, the JI station was blacked out for the rest of the night.

In the Australia and New Zealand area (Sidney, Aukland, Melbourne), many MF and HF transmissions were strongly attenuated; signal strengths were down an average of 30 db during the first hour, but improved after H + 60 minutes. There was also strong attenuation of radio signals from Honolulu. For instance, Voice of America directed from Honolulu to New Zealand and Australia on 9.65 MHz was down 30 db at H + 5 minutes, and down 20 db at H + 60 minutes. "Enhanced D-region ionization continued to be apparent for the remainder of the night, as no distant MF station or HF station below 20 MHz returned to its normal nighttime signal strength."

At Rarotonga, similar effects were observed at MF and in the lower HF range. During the following nights, New Zealand and Australian MF stations faded out completely, but not the US Stations.

BBC transmissions to Wellington, New Zealand, were again enhanced in the 15- to 21-MHz range.

Worldwide Effects. Observations of mostly transient effects of VLF transmissions were reported from the State of Washington; Boulder, Colorado; Panama; Chile; Wellington, New Zealand; and many other places. It does not appear that the transient effects posed a serious communication problem, although it would seem worth while to make a more comprehensive study of the exact physical sources for these perturbations. Speculations have been advanced that some of these effects were produced by neutron-decay protons and electrons.[22]

An increase in the absorption of 30-MHz cosmic radio noise was observed at four stations in Alaska within two seconds after the explosion, by Basler, Dyce, and Leinbach.[23] The authors believe that the ionization in the lower ionosphere originated at the endpoints of the radiation belt tubes formed at L = 1.5 to L = 2.0.

Other Events

Checkmate. HF interference was serious in the JI area for approximately half an hour. There was strong-to-moderate HF interference in the Pacific area to distances of ~1 000 km from JI for one or two hours.

Kingfish. The communication disruptions were widespread and moderately severe. Actual HF communications to and from JI were out for about three hours. Ionosonde measurements implied complete blackout of HF frequencies for at least one hour due to heavy absorption in the D-layer.

There is little information in the unclassified literature, presumably because yield and altitude are still classified. Interpretation of observations is therefore ambiguous.

Bluegill. Degradation of communications was relatively moderate (compared with Teak, Orange, and Starfish). It was predominantly local. HF on JI was out for about two hours. Effects at larger distances were generally small or minimal. Many details are in the classified literature.

Geomagnetic worldwide effects were, by many orders of magnitude, smaller than for Starfish—understandably so, because at low altitude, the particle pressure is the dominant factor in fireball phenomenology.

Tightrope. Very moderate southern conjugate ionospheric effects occurred. HF Midway-Palmyra links were not affected (they pass in JI vicinity). There are some details on communication interference during the first hour after the event in the classified literature. Generally, they were small.

Supplementary Information

It is worthwhile to have a sharper look at the reasons why the communications interference in the Hawaii area was delayed after the Teak event and more so after Orange.

Let us take Teak, for example. The prompt gamma-ray output was high, nominally 0.2% of the yield. The arc from the burst point to the D-layer at ~50 km above Honolulu is about 1 200 km long; the shortest approach of the chord to the surface of the earth occurs at an altitude of 20-25 km. Consequently, the gamma rays had to penetrate about 7 air masses and were attenuated by a factor of 1 000. Still, the gamma flux is strong enough to generate an electron density of the order of 10^8 electrons/cm^3, but only for microseconds. The electrons are removed very quickly by attachment to O_2 and more complex compounds. This very transient increase in ionization was probably unobservable by commercial equipment.

If we look now at the much more steady flux of fission-product beta-rays,[24] say at plus one second after Teak burst time, then the electron density above Honolulu increases only slightly—namely, by 5 electrons/cm^3 at 50 km and by 15 electrons/cm^3 at 70 km. Thus, the density increase in the D-layer is of the order of only 10%.

However, as time goes on, the debris cloud rises. At plus 5 minutes it reaches 400 km,[25] and at about plus 20 minutes, it is 500 km above JI with a horizontal dimension of the same magnitude. While the gamma activity has decreased substantially, the transmission to the air above Honolulu has increased by almost an order of magnitude. Furthermore, the penetration into the D-layer increases, and so does the column electron density because of the steeper look-angle of ~20°. This leads to electron-density increases of one to two orders of magnitude* above ambient at this time, consistent with the radio-frequency observations.

Concluding Remarks

The communication interference picture as presented in this report is not complete. In a few instances, attempts have been made to associate specific interferences or blackouts with the source characteristics, i.e., burst location, debris motion, prompt and delayed radiations, their attenuation, etc. It would be desirable to present a still better, fully coherent story of the whole pertinent phenomenology. While today's knowledge of the *late* phenomenology could conceivably be improved by putting more bits and pieces together—a tedious task—the full picture would probably not evolve, simply because of limitations in observational data. Furthermore, the theoretical treatment of these late phases of the phenomenology and of the atmospheric interactions is difficult to do with confidence. Nevertheless, the information gained so far is of great qualitative and semiquantitative value.

———————————

*Critical electron densities, electrons/cm^3:

30 kHz	10^1
300 kHz	10^3
3 MHz	10^5
30 MHz	10^7
300 MHz	10^9

CHAPTER V

THE FORMATION OF ARTIFICIAL RADIATION BELTS.
EFFECTS ON SATELLITES.

Argus

Before the discovery of the natural Van Allen belts in 1958, N. C. Christofilos[26-28] had suggested in October 1957 that many observable geophysical effects could be produced by a nuclear explosion at high altitude in the upper atmosphere. This suggestion was reduced to practice with the sponsorship of the Advanced Research Project Agency (ARPA) of the Department of Defense and under the overall direction of Herbert York, who was then Chief Scientist of ARPA. "It required only four months from the time it was decided to proceed with the tests until the first bomb was exploded." The code name of the project was "Argus." Three events took place in the South Atlantic. Data, yields, and locations are shown in Table I. Following these events, artificial belts of trapped radiation were observed.

A general description of trapped radiation is as follows. Charged particles move in spirals around magnetic-field lines. The pitch angle (the angle between the direction of the motion of the particle and direction of the field line) has a low value at the equator and increases while the particle moves down a field line in the direction where the magnetic field strength increases. When the pitch angle becomes 90°, the particle must move in the other direction, up the field lines, until the process repeats itself at the other end. The particle is continuously reflected at the two "mirror" points—it is trapped in the field. Because of asymmetries in the field, the particles also drift around the earth, electrons towards the east. Thus, they form a shell around the earth similar in shape to the surface formed by a field line rotated around the magnetic dipole axis. The shells are called L-shells; the L-value is the ratio of the distance of the equatorial crossing point of the field line from the center of the dipole to the earth's radius. (In reality, the dipole field is somewhat distorted.) The approximate L-values of the 1958 detonations were 1.7, 2.1, and 2.0 for Argus I, II, and III respectively.

The rockets carrying the nuclear devices were launched from shipboard. Measurements were made by Explorer IV.[28] Additional Argus II data were obtained by sounding rockets.[29]

The artificial belts formed between $L = 1.7$ and $L = 2.2$. This is between the inner and outer zones of the natural Van Allen belts. The center of the inner natural zone is between $L = 1.15$ and $L = 1.3$; a broad slot of low intensity is located at about $L = 2.8$, and the center of the outer zone is near $L = 4.5$. Because of the presence of energetic protons, relatively little was known before 1962 about the omnidirectional electron populations in the inner zone. Hess[30] gives the following numbers for the 1957 fluxes: $E > 40$ keV: 3×10^7 electrons/cm²-s; $E > 580$ keV: 2×10^6 electrons/cm²-s. Numbers are uncertain at least by a factor of three. The fluxes in the slot region are three to four orders of magnitude lower than those in the inner zone. The original Argus data were published in terms of count rates; they imply that the fluxes were about one order of magnitude larger than the natural flux densities in the respective natural shells. Later, Van Allen[31] gave maximum omnidirectional fluxes of 10^5/cm²-s for Argus I and II, and 10^6/cm²-s for Argus III. These artificial belts were stable for several weeks; belts I and II were 90 km thick, and belt III was 150 km thick. The electron lifetimes—i.e., the time for the electron fluxes to decrease by factor e—were 6-10 days for electron energies > 3 MeV.[32] Injection efficiencies were difficult to derive from Explorer IV data. Estimates vary from 12 to

27% for low energies and from 2 to 11% for $E > 5$ MeV;[32] the authors advise caution in the use of these numbers. For further details, see the review by Cladis, Davidson, and Newkirk in the "Trapped Radiation Handbook."[32]

Unfortunately, the optical photographic coverage of the events was inadequate. Only recently, some rather scarce photographic records obtained on Argus II and III were analyzed, and the phenomenology was found to be similar to that observed on the Starfish event.[33,34] On Argus III, an electron patch formed at about 65 km and a debris patch near 100 km altitude in the conjugate area; later on, field-aligned striated ionization was seen at higher altitudes.

The Argus experiments were originally kept classified, but their occurrence and results were later made public[35] because of limited military significance. The purely scientific results were of greater value. The artificial belts had low electron fluxes, and they did not interfere with the study of natural phenomena. At least I am not aware of any critical complaints about undesirable environmental effects.

Teak and Orange

The study of Explorer IV records by Johnson and Dyce[36] after the Teak and Orange events provided evidence for trapped radiation in both instances. The Teak belts were more pronounced; they lasted for several days and centered at $L = 1.2$. Radar backscatter data[36] indicate that five hours after Teak, the debris cloud of fission products was centered some 600 km west-northwest of JI and that it had dimensions of several hundred kilometers at altitudes between 100 and 200 km. $L = 1.1$ to $L = 1.2$ would intersect this debris cloud at approximately 200 km. Van Allen[31] gives 10^9/cm²-s as the maximum omnidirectional fluxes for Teak and Orange. The total number of all electrons trapped in the Teak shell at plus one hour is estimated to have been $\sim 10^{20}$, indicating a very low injection efficiency of about 10^{-7}. For more details, refer to the "Trapped Radiation Handbook."[32]

Starfish

This event produced by far the most intense, long-lasting radiation belt. At burst time, several satellites were in rather low orbits; their apogees were near 1 000 km. At D + 1 day, Telstar was launched into a more favorable elliptical orbit which covered the space up to an apogee of 5 600 km. Subsequently, several other satellites were launched which provided additional data. For a listing, see the "Trapped Radiation Handbook,"[32] Sections 6-33.

Because of differences in orbits, spectral coverage, and launch times, the data obtained by the various satellites did not always agree. However, the maximum electron fluxes were encountered between $L = 1.2$ and $L = 1.4$; at D + 1-2 days they amounted to $\sim 10^9$ electrons/cm²-s. In these L-shells, the spectrum was similar to the fission spectrum. At higher altitudes and L-shells, significant discrepancies between the observations of the satellites in orbit before explosion time[37-39] and of the post-burst Telstar[40] became the subject of many scientific arguments. The Telstar instrumentation had better spectral resolution, and the measurements extended to high L-values where the fluxes observed were much higher and softer than were those indirectly derived from the other satellites, which entered high L-space only at high latitudes. The validity of the interpretation of these Telstar data as owing to Starfish electrons was questioned, because no pre-event data existed for this part of the space. The high population could have been due to natural causes, or quite possibly these low-energy electrons may have been injected from shock-heated air in the exosphere as postulated by S. Colgate.[41] Even today the problem has not been fully resolved. As time went by, the differences in terms of the total electron inventory narrowed, however. By 1966, the following picture evolved: Van Allen[31] assumed that the nominal yield of fission-decay

24

electrons was 5×10^{26}. This is reasonably close to my number of 7.5×10^{26} derived from Griffin.[42] The reported inventories at $\sim$D + 10 hours were

O'Brien et al., Injun I, 1962[37]	10^{24}
Van Allen, 3 satellites, 1965[31]	1.3×10^{25}
Hess et al., Telstar, 1962[40]	2×10^{26}
Hess, Brown, and Gabbe[40]	7×10^{25}
(Walt and Newkirk, 1971[32]	7.5×10^{25})

A probable mean number is 4×10^{25}, corresponding to an injection efficiency of $\sim$5%. It is most likely that the injection occurred by way of the strong debris-jets moving across the magnetic field lines as observed from Christmas Island[43] and by a high flying aircraft.[44] Debris were observed photographically at altitudes up to 2 000 km and at times between one and three minutes after the burst (Figs. 6,7).

The decay of the electron population has been treated in considerable detail. Van Allen[31] reports that 15% of the total injected survived 5-1/2 months and 10% survived 12 months. The lifetimes are dependent on electron energies, the shell, and the B-field. Low-energy electrons have shorter lifetimes. Thus, a fission spectrum becomes harder as time goes by.

Decay at low altitudes is caused by scattering interactions with air; this decay is fast, roughly about 50 days at 400 km. Decay in the main belt was found to be of the order of about three months to several years. The decay is caused by coulomb scattering with atmospheric atoms. The decay at higher altitudes (L > 1.7) cannot be caused by collisions with atmospheric constituents. It is believed to be caused by magnetic disturbances, i.e., by interactions with solar-wind-induced electromagnetic waves. While the lifetime at L = 1.7 is many months, the decay time at L = 2.2 is of the order of one week.

Neutron Decay Electrons. Hess[30] reviewed the contribution of Starfish neutrons to the radiation belt. Neutron half-life is about 1 000 seconds, with decay into protons and electrons. Calculations lead to numbers of the order of 10^6 to 10^7 trapped electrons/cm²-s in the main Starfish belt. This number is not negligible, but is much lower than the originally trapped fission-electron fluxes. Some effect on VLF propagation is indicated.

Checkmate, Kingfish, Bluegill

Although some air fluorescence was observed in the southern conjugate area after all three events, and synchrotron radiation was measured along the magnetic meridian through JI after Checkmate for a short time, there seems to be no evidence for the formation of artificial radiation belts of significant lifetime.

Satellite Damage from Starfish Electrons

Ariel. US-UK satellite Ariel was launched from Cape Canaveral on April 26, 1962. Orbit inclination was 54°, apogee 1 209 km, and perigee 393 km. At Starfish explosion time, Ariel was at a distance of 7 400 km from JI. After July 13, 1962, four days after the explosion, Ariel operated only intermittently as a result of the deterioration of the solar cells owing to the effects of the artificial radiation belts.[45]

25

Traac. Traac, the research satellite of the Applied Physics Laboratories, Johns Hopkins University, had operated for 190 days at the time of the Starfish event. The solar-cell power system had already suffered some degradation in the natural space environment. Traac's apogee was 1 100 km, perigee 951 km, inclination 32°. Data were received on a reduced schedule between days 190 through 224, because of accelerated degradation of solar cells. After day 224, i.e., 34 days after the explosion, the satellite failed to transmit data.[45,46]

Transit 4B. W. N. Hess[40] reports that "...on August 2, Transit 4B stopped transmitting...." In Table I of the same publication, Traac and Transit 4B are shown to have the same orbit, i.e., 960 km perigee and 1 106 km apogee and an inclination of 32°. The solar-cell damage curves in his Fig. 8 are shown to be the same for Transit,[40] Traac, and Ariel I. However, Ariel started to have trouble at plus 3-1/2 days (different orbit, though encountering lower fluxes than the other two), Transit at plus 25 days, and Traac at plus 38 days. Hess[30] has a listing supposedly of all satellites launched before July 1, 1966, but Transit 4B is not listed.

Cosmos V. Cosmos V was launched in Russia on May 28, 1962. Orbit inclination was 49°; apogee on July 9 was 1 512 km, perigee 204 km. The article describing results contains only indirect information on the lifetime of power sources of the satellite.[47] Radiation belt data were taken over a period of four months. The design of the satellite is described in an article by V. I. Krassovsky et al.[48]

Injun I, Telstar. Injun I's active life ended in December 1962—apogee 1 020 km, perigee 860 km, inclination 67°. Telstar transmitted through February 1963—apogee 5 600 km, perigee 955 km, inclination 45°. I have been told that Telstar developed some component trouble which, however, could be overcome by command to a back-up circuitry, after ^{60}Co irradiation of a mock-up assembly.

The vulnerability of solar cells and electronic circuit components to nuclear radiations has been treated extensively in the literature. References 32 and 40 are pertinent.

Effects on Manned Spacecraft

This is a complicated subject; the dose received by an astronaut depends on many variables such as type of orbit (apogee, perigee, inclination), degree of shielding, and duration of flights. It will be very different for Gemini or Skylab-type flights, which would be seriously affected by artificial belts, and, on the other hand, for Apollo flights, which are subject only to transient radiation-belt effects, to solar-wind, solar-flare, and cosmic radiation.

Adams and Mar[49] provided daily dose rates from electrons, protons, and bremsstrahlung for various orbits and for shielding by 0.4 and 2.0 g/cm² of aluminum. They also refer to experimental dose measurements made in unmanned satellites as follows:

Peak readings four months after the Starfish explosion occurred at B = 0.16 and L = 1.25 (1 600 km above the equator). They were as follows:

- 30 rads/h in a chamber shielded by 4.7 g/cm² brass,
- 23 000 rads/h in a chamber shielded by 0.4 g/cm² brass.

Thus, for a satellite in a polar circular earth orbit, the *daily* dose would have been at the very least 60 rads in a heavily shielded vehicle at Starfish time plus four months. A corresponding maximum dose rate of 0.15 rads/h was measured before the Starfish event by the heavily shielded chamber in almost the same location (B = 0.2, L = 1.25).

One can also calculate the dose an astronaut in a Skylab-type orbit (roughly circular at 435 km, 50° inclination, 90 minutes per orbit) would receive, say at one week after Starfish. The maximum dose would amount to ~50 rads per orbit behind 1 g/cm² and ~5 rads behind 2 g/cm² of aluminum. This is about 1 000 times the natural dose. The average yearly dose limit

26

recommended by NASA[32a] for eyes is 27 rads (the eyes are the most sensitive part of the body). Lifetime total body exposures of 400 rads of penetrating radiation for early space explorers and up to 200 rads for future space passengers have been suggested as compatible with a reasonable risk.[50] I have no numbers for Skylab shielding. The Gemini spacecraft shielding varied from 1.6 g/cm² to 7 g/cm² of aluminum, depending on the solid angle subtended by the exposed body areas.

For more details, reference is made to the "Trapped Radiation Handbook,"[32] to "Status Report on the Space Radiation Effects on the Apollo Mission,"[32a] and to "Radiation Trapped in the Earth's Magnetic Field."[51] In addition, W. H. Langham[31b] was leading an extensive study of radiobiological factors in manned space flight for the National Academy of Sciences.

Russian Events

The Russians conducted three high-altitude tests in October and November 1962 at high L-values. Van Allen[31] provided the following data:

Date of Burst	Nominal Yield	Maximum Omnidirectional Intensity at t=0	L-Value of Burst	Approximate Mean Lifetime
22 Oct. 1962	Submegaton	10^7 e/cm²-s	1.9	1 month
28 Oct. 1962	Submegaton	10^7 e/cm²-s	2.0	1 month
1 Nov. 1962	Megaton	10^7 e/cm²-s	1.8	1 month

The maximum fluxes of the Soviet belts are two orders of magnitude lower than those encountered in the Starfish belt, and are one to two orders of magnitude higher than in the Argus belt. The short mean lifetime is probably owing to pitch-angle scattering and to loss at the mirror points.

Response of the Scientific Community

The response was mixed. This subject matter is reviewed separately in Chapter XI.

CHAPTER VI

SYNCHROTRON RADIATION AND HYDROMAGNETIC WAVES.

Electrons moving in a circular orbit perpendicular to a magnetic field will be accelerated and, consequently, emit radiation. Low-energy electrons radiate at the frequency of the circular motion; this frequency is often called the cyclotron frequency. High-energy relativistic electrons emit radiation in the direction of the motion of the particle at frequencies higher than the cyclotron frequency. This is called synchrotron radiation or, sometimes, magnetic bremsstrahlung.

Synchrotron radiation was observed after Starfish at Central Pacific sites with existing riometer networks operating in the 30- to 120-MHz range.[52] The best data were obtained in Huancayo, Peru, where a strong burst of radiation was observed promptly after the Starfish explosion. "Beginning ten minutes later, the original intense tube of emitting electrons appeared again over the Central Pacific, having completed one trip around the globe...." Synchrotron radiation was also observed at Canton and Palmyra Islands south of JI, but the onset was delayed; in fact, the natural cosmic noise was first reduced in these areas by D-region attenuation caused by prompt x-rays. Wake Island, about 2 500 km west of JI, observed synchrotron radiation at plus 10 minutes after the burst. The radiation was strongest at the magnetic equator: the intensity fell to half maximum at 12° and to one-tenth maximum at 20° from the equator. The effect was difficult to detect at radio astronomy stations at higher latitudes. The intensity decayed slowly with a decay constant of about 100 days, in general agreement with satellite data quoted earlier. The initial electron spectrum corresponded to a fission spectrum.

It was concluded[53] that the radiation is generated primarily at heights above the equator between $L = 1.2$ and $L = 1.6$. Attempts were also made to derive the total number of electrons with energies >1 MeV that were trapped after the explosion.[54]

Little is known experimentally about the synchrotron noise produced at lower frequency, although a British radio astronomer[55] conjectures that the artificial radiation "...may be a hazard to accurate work on the galactic spectrum below 25 MHz."

Magnetic Disturbances and Hydromagnetic Wave Observations

After the Starfish detonation, changes in magnetic field strength and geomagnetic micropulsations were observed at many stations in North and South America, South Africa, Australia, and Japan. The changes were of a transient nature; their onset varied from about plus two to five seconds. The interpretation of the physical processes which caused the disturbances is still ambiguous.

One school of thought associates the onset of the perturbations with the arrival of an MHD wave at the particular location. Magnetic containment of bomb debris results in the stretching of the magnetic field lines; the stretching of the lines is propagated as a transverse MHD wave along the B-lines. Therefore, a zero MHD effect would indicate no containment in the early phases of the detonation. It is well known[45] that the initial expansion of the Starfish debris was contained by the magnetic field. A manifestation of this containment was observed at three geophysical stations in Peru at distances of 9 000 km from the burst.[56] The magnetic activity at these stations was interpreted as being caused by the arrival of an MHD wave traveling at a velocity of 5 x 10^8 cm/s at a height of ~1 000 km. Furthermore, the magnetic field lines above Huancayo in Peru were sufficiently distorted for a few minutes to permit an increase in the cosmic-ray flux at the station.

Somewhat similar interpretations were advanced by Stanford University and California Institute of Technology[57] researchers. It is stated that "the Alfvén velocity-altitude profile indicates that MHD waves can be trapped in a waveguide in the altitude range from 300 to 3 000 km, thus producing resonant oscillations."

The coincidence of geomagnetic field fluctuations and variations in E-layer ionization density is advanced by another group[58] as evidence that the perturbations were caused by interactions of nuclear radiations with the magnetic field and ionosphere. Actually, the E-layer could be "seen" at most of the reporting stations by the debris jets and by the higher L-shells.

Another interpretation was advanced by Nawrocki,[59] who calculated the size of similar Argus signals and the propagation time in terms of the diamagnetic effect of the thermal electrons which spiral along the B-lines and give rise to a decrease in the ambient field. However, there is no reason to believe that the Argus debris expansion was not magnetically contained initially.

In any case, while observed worldwide, the magnetic disturbances did not seem to interfere with the normal activities of the geophysical stations; they may have contributed to a better understanding of somewhat similar natural phenomena, such as the "sudden commencements."

30

CHAPTER VII

ELECTROMAGNETIC-RADIATION EFFECTS ON
ELECTRICAL AND ELECTRONIC SYSTEMS.
(Contribution by John Malik)

The electromagnetic radiation in the radio-frequency portion of the spectrum (EMP) can cause problems in electronic systems. The pulse from detonations above about 30 km is caused by the deflection of Compton electrons produced by gamma-ray interaction with the earth's magnetic field in the deposition region (20-40 km). The resulting transverse current in the large area of gamma-ray deposition produces a large coherent radiating element. With appropriate yield, detonation altitude, and magnetic azimuth, the electric fields over large areas at the earth's surface can exceed 10^4 V/m. Such fields can cause detrimental effects on some types of electrical systems. The pulse width is less than a microsecond.

Starfish produced the largest fields of the high-altitude detonations; they caused outages of the series-connected street-lighting systems of Oahu (Hawaii), probable failure of a microwave repeating station on Kauai, failure of the input stages of ionospheric sounders and damage to rectifiers in communication receivers. Other than the failure of the microwave link, no problem was noted in the telephone system. No failure was noted in the telemetry systems used for data transmission on board the many instrumentation rockets.

There was no apparent increase in radio or television repairs subsequent to any of the JI detonations. The failures observed were generally in the unprotected input stages of receivers or in rectifiers of electronic equipment; transients on the power line probably caused the rectifier failures. There was one failure in the unprotected part of an electronic system of the LASL Optical Station on top of Mount Haleakala on Maui Island. With the increase of solid-state circuitry over the vacuum-tube technology of 1962, the susceptibility of electronic equipment will be higher, and the probability of more problems for future detonations will be greater. However, if detonations are below line-of-sight, the fields and therefore system problems will be much smaller.

31

CHAPTER VIII

AURORAL PHENOMENA.
DETECTION OF NUCLEAR EXPLOSIONS IN SPACE.
HYPOTHETICAL EFFECTS ON WEATHER PATTERNS OF ENERGY DEPOSITION IN THE UPPER ATMOSPHERE.

General

The interaction of nuclear radiations—gamma rays, neutrons, beta particles, x-rays—and of debris with the surrounding air molecules produces quantities of visible light and ultraviolet and infrared radiation. Gamma-ray-excited air fluorescence was already observed at Trinity. Most of the radiation is emitted from excited nitrogen, more specifically by the second and first positive systems of N_2 and the first negative and the Meinel systems of N_2^+. The efficiencies for conversion of nuclear emissions into light is low at sea level because of collisional deactivation of the excited states; it is much higher at lower air densities.[60-62] Consequently, at higher altitudes, all fluorescence emissions are more brilliant[63] and quite spectacular indeed. They have many similarities with natural auroral phenomena, where the same emissions, excited mainly by auroral electrons, are present.[64] To my knowledge, studies of manmade aurorae did not interfere with normal auroral research activities.

Teak

The Teak event provided the most impressive display; as seen from JI, it was breathtaking. Excellent documentation was obtained from Mt. Haleakala on Maui by J. Champeny of EG&G, as reported in Ref. 25 (see Fig. 3). An auroral streamer originating from the debris mass was clearly visible, moving upwards along the field lines towards the equator. The northern branch of the aurora was observed from JI and from two aircraft.[25] The Maui pictures show the rise of the luminous debris to about 500 km and their late field-aligned structure; also seen are the luminous-air shock, the debris-air shock, and the debris. The slowing down of the upward-moving shock between altitudes of 300 to 500 km was interpreted as being caused by work against the magnetic field, with energy being carried away by MHD waves. The auroral streamer moving south deposited its energy at the conjugate point were bright auroral arcs were seen from Apia as reported by Matsushita,[65] who published a review of many geophysical effects generated by the Teak and Orange events. The auroral-type displays after Orange were less spectacular because of the lower burst altitude.

Starfish

The Starfish aurorae were somewhat less brilliant than those following Teak; they were seen, however, over a larger area. The x-ray-produced fluorescence at the stopping altitude of ~100 km was of very short duration. The surface brightness was not very high, but the total emitting area directly underneath was in excess of 100 000 square kilometers. Longer lasting luminous patches of air excited by debris and fission-product beta rays occurred in the northern and southern conjugate areas. The displays in the south were observed at some 20

33

geophysical and meteorological stations from the equator (Tarawa) down to Apia, in many places in New Zealand, and even at Campbell Island (53°S).[66] Details of all the geophysical effects were published in 21 papers of a Special Nuclear Explosion Issue of the "New Zealand Journal of Geology and Geophysics."[67] The most interesting result of this survey is the very large extent of the affected area: debris and debris electrons deposited their energies over some 40° of latitude in the south (Tarawa observed the luminosity in the overhead tube). These observations supplement the air-based gamma-ray measurements reported by D'Arcy and Colgate.[21] These auroral observations, together with others taken from airplanes, from Hawaii, and from Christmas Island, were invaluable in deriving a general physical picture of the Starfish phenomenology.[43] For instance, the optical observations of jets of excited plasma directed initially along the field lines, but at higher altitudes moving straight across the field (Fig. 7), provided the information needed to fully understand the formation of the artificial radiation belts and the ionization and luminosities produced far to the south of the conjugate area. Starfish was indeed a large-scale demonstration of many principles of plasma and auroral physics. The results confirmed the anticipation of the scientific usefulness of nuclear explosions in space as expressed in 1959 by LASL staff.[68]

Checkmate, Kingfish, Bluegill

Auroral displays occurred in all these events on a smaller scale, both in the burst area and at the southern conjugate locations.

Detection of Nuclear Explosions in Space

During the moratorium between the 1958 and the 1962 test series, the auroral data obtained on the Teak event were applied to the design and construction of the Los Alamos Air Fluorescence Detection System for possible clandestine nuclear explosions in space. The range for detection was $R = 10^6 \sqrt{Y_x}$ kilometers in daylight, where Y_x is the thermal x-ray yield of the explosion in kilotons.[69]

Effects on Weather Patterns: Link Between Magnetic and Atmospheric Storms

Generally there has been a great deal of conversation about effects of nuclear explosions on weather patterns. However, I am not aware of pertinent serious studies, although I have not made a thorough search of the literature. As to the high-altitude events (and of these, especially the high-yield explosions), again no information seems to exist. This is not surprising, since the latent heat in large air masses is much larger than the energy release in a bomb. On the other hand, recent studies of a possible relationship between certain auroral displays in the north and weather do not exclude the hypothetical possibility of artificial weather-modification by nuclear-energy releases.

W. Orr Roberts and R. H. Olson appear to have confirmed a statistical relationship between the behavior of low-pressure troughs over the Gulf of Alaska preceded by northern lights and the behavior of those formed and moving in their absence.[70,71] They report that during winter, low-pressure troughs tend to intensify or deepen, in response to storms in the geomagnetic field which produce auroras. It is well known that the development of low-pressure areas in the Gulf of Alaska has a strong influence on North America's weather. About one-third of these low-pressure systems move into the central United States. Now, those preceded by northern lights are reported to penetrate about 200 miles farther south and to bring colder weather with them. "...Although not all large troughs are triggered by northern

34

lights and not all auroras are followed by trough development, the probability of the trough's intensifying seems to be approximately doubled by the occurrence of a magnetic storm." More recently the studies by Roberts, Olson, Wilcox, and others[72] were extended to the whole Northern Hemisphere. They relate the vorticity area index, essentially a measure in square kilometers of the size and prominence of all low-pressure troughs in the Northern Hemisphere north of 20°N, to weather patterns. The vorticity area index is affected both by the rather regular sweep past the Earth of the interplanetary magnetic-field sector boundaries of solar origin and by the more irregular occurrences and magnitudes of solar flares. The maximum of the local vorticity area index for a sector 60° wide in longitude (i.e., more than the width of the Gulf of Alaska) is 2×10^6 km².

While the data collected so far cannot be neglected, the physical explanation of the statistical relationships is the subject of much speculation.

In this context it is interesting to compare the energy depositions from the precipitation of natural electrons in an area approximately 500 km in latitude and 2 000 km in longitude with the energy deposited by a bomb. Typical auroral fluxes range from 1 to 100 erg/cm²-s. Assuming a three-hour display and the highest flux, the total energy deposition in this area would be 10^{22} erg = 1/4 Mt. Chamberlin reports bombardment energies of as much as ~400 ergs/cm²-s in a bright aurora (Ref. 64, p. 28); this is equivalent to 1 Mt incident over 10^6 km².

The Starfish x-ray energy deposition over a radius of 300 km was about 1/4 Mt, although the pulse duration was much smaller. Conceivably, a few events like Starfish over the Gulf of Alaska could add to the statistical studies referred to above. International political problems and environmental-impact considerations would, however, pose almost insurmountable obstacles to such an exercise, although the use of relatively clean sources, the long residence time of the debris in the stratosphere, and advance knowledge of communication problems would soften the impact. The effects on satellites, both by prompt radiation and by short-lived though weak radiation belts, would be most difficult to prevent or overcome.

The study of the coupling processes between the thermosphere where auroral particles and x-rays are stopped and the mesosphere, stratosphere, and the upper troposphere—i.e., the meteorologically important 300-millibar, ~9-km-altitude range, remains a most interesting unsolved problem of upper-atmospheric physics. Direct injection of condensation nuclei and/or water vapor into the latter altitude domain would also need to be considered. In this context, reference is made to recent experimental feasibility studies and speculations about the initiation of large-scale atmospheric motion by other physical means, such as intense radio waves, TNT explosions, the kinetic energy obtained by the fall of tons of material into the upper atmosphere, etc.[73]

In any case, the relevance of nuclear-explosion effects to weather modification problems requires much more critical discussion than is attempted in this paper.

CHAPTER IX

RESIDENCE TIME OF RADIOACTIVE TRACERS IN THE STRATOSPHERE. UPPER ATMOSPHERIC AIR-CIRCULATION PATTERNS.

Injection of ^{102}Rh and ^{109}Cd

The Orange and Starfish warheads contained special tracer elements created by neutron activation from the devices. About 3 megacuries (MCi)* of ^{102}Rh were produced in the Orange weapon;[74] this isomer of principal concern has a half-life of 210 days. The main debris mass rose to an estimated altitude of 150 km, although relatively crude observations from Mt. Haleakala on Maui Island indicated some debris as high as 500 km.

In the Starfish tracer experiment, 0.25 ± 0.15 MCi of ^{109}Cd were produced.[75] Cadmium-109 has a half-life of 470 days. The Starfish explosion occurred 400 km above JI. As discussed elsewhere in this report, approximately 30% of the debris were initially deposited in the general burst area; about 30% each were guided by the magnetic field to the northern and southern conjugate areas, where debris patches were formed at altitudes of 100-120 km. The heating of the air led to a subsequent rise of the patches. The balance of the debris were ejected to altitudes of 2 000 km or more. Thus, the major debris masses were initially distributed evenly in both hemispheres at altitudes between 100 and 200 km.

Observations

Information on the distribution of the debris in the lower stratosphere was obtained from balloon data up to 28 km, from aircraft observations up to 19.4 km, and from surface measurements. Kalkstein,[74] List and Telegadas,[76,77] and others have studied the time and space history of tracer motion in considerable detail.

Rhodium-102 was first observed in the south in small quantities. Larger, significant concentrations were measured in May 1959 near 19.4 km altitude between 45° and 60° south. The onset of similar concentrations at northern latitudes, same altitudes, occurred four months later in September 1959, indicating an initial movement of the debris towards the winter hemisphere; maximum concentrations were measured in the north beginning in February 1960, remaining constant throughout 1961. In the south, about the same maxima were reached later, in May 1960, and also remained constant throughout 1961. During this period, the low-latitude inventories were approximately two to three times lower than those at high latitudes.

The vertical motion of the ^{102}Rh tracers was faster at high latitudes than at low latitudes. The first measurable, though small, quantities in the north were obtained at 28 km, in March 1959; by June 1960, the tracer particles had moved to the lower stratosphere at 70° north and approximately 12.5 km altitude, whereas at 25° north they had descended more slowly to a height of 20 km.

---

*One curie (Ci) corresponds to 3.7 x 10^{10} disintegrations/s.

The global inventory of [102]Rh as of May 1961 (i.e., 33 months after the injection) was reported[76] to be as follows, in MCi corrected for decay to August 12, 1958:

Deposition	~0.02
Troposphere	~0.05
Stratosphere to 21 km	~0.48
Stratosphere 21-31 km	~0.50
Unaccounted	~2.00

Cadmium-109 was first observed by the AEC's balloon-sampling program in December 1962 at 35 km altitude in the south and several months later in the north.[75] On the whole, the observations confirmed the results suggested by the [102]Rh tracer motion; but in this case, the concentrations measured at high latitudes both in the north and the south were up to 10 times higher than in the equatorial areas. Latitudinal cross sections of mean seasonal [109]Cd stratospheric concentrations as a function of altitude for the period December 1962 through August 1966 were published by Telegedas et al.[77,79,80] Among others, a group of Russian workers[81] measured [109]Cd fallout on the ground with large collectors at four places in the Soviet Union (Moscow, Tbilisi, Vladivostok, and Arkhangelsk) during 1964-1967.

Residence Time

It is perhaps well to begin by defining "residence time." Most authors are rather lax in their use of this term. Mean stratospheric residence time is normally defined as "the average time spent by radioactive debris in the stratosphere" before it is transferred to the troposphere. I presume that this is the time for concentration to drop by factor e. Half-residence time is also sometimes used. The mean residence time in the troposphere is generally taken as 30 days; removal is mainly by rain-out.

Kalkstein[74] derived a high-altitude [102]Rh tracer "residence" time of "roughly ten years." Volchok[82] assumed a model atmosphere above the tropopause consisting of two atmospheric layers. On the basis of [102]Rh, the region above 21 km has a *half-time* for removal of 10 years; the removal half-time from the lower stratosphere is taken as 2 years. On the basis of the [109]Cd data, Volchok obtained a similar model for the period of 2-3 years after the explosion, again 10 years for the half-time in the upper stratosphere but only 1 year for the lower stratosphere. The Russian workers Leipunskii et al.[81] generally agreed with the US interpretation and concluded that "finely divided aerosols injected above 100 km are removed from the upper atmosphere with a half-time of about ten years and a mean residence time of fourteen years."

It is important to make a clear distinction between injection at altitudes of >100 km by high-altitude explosions and injection into the lower stratosphere by cloudrise from megaton explosions at or near the surface. In the latter case, the residence times of fission products in the lower stratosphere are much shorter than for those transported to altitudes of >100 km. The debris clouds of megaton-size explosions rise to altitudes of the order of 20 km and penetrate the tropopause into the lower stratosphere. The UN document of 1972[83] quotes the following residence times:

- Lower Polar Stratosphere, 6 months
- [90]Sr, Lower Stratosphere, 1-1.2 years.

The fallout characteristics of ^{90}Sr, half-life 27.7 years, have been extensively studied because of its biological significance. The numbers differ somewhat from author to author; for instance, for low-altitude bursts, Fabian, Libby, and Palmer[84] obtained a stratospheric

38

residence time of 1.6 years. Also for low-altitude, high-yield bursts, Peterson[85] reports half-residence times of five months for injections into the lower polar stratosphere and two years for injections into the "upper polar stratosphere," the latter from the Russian 1961 test series.*

I believe it is safe to say that residence times of debris injected into the lower stratosphere are of the order of 1-2 years, in contrast to residence times of 14 years for injections at 100 km or more.

Stratospheric Circulation

The first generalized model of the circulation of the [102]Rh tracer was developed by Kalkstein[74] in 1962. Stebbins[86] confirmed, as a result of the DOD's High Altitude Sampling Program (HASP), Kalkstein's AEC Health and Safety Laboratory (HASL) data. He agrees with Kalkstein, namely that "it appears that the [102]Rh is being brought down from its original injection site by strong vertical mixing in the polar stratosphere during the winter season.... The [102]Rh in the tropical atmosphere probably reached there through downward mixing in the polar stratosphere and then lateral mixing to the tropical stratosphere." Stebbins also reports that particles as small as 0.001 micrometers in diameter fall from 300 km to 80 km in a matter of days and that the fall rates begin to decrease markedly once these particles reach the denser air of the mesosphere. Subsequently, Telegadas and List,[76] also generally agreeing with Kalkstein and Stebbins, suggest that debris injected at very great heights over the equatorial region descend into the polar stratosphere and are subsequently propagated downward and equatorward. They find that north of 35°N, between 14 and 20 km, the downward movement in the winter months is of the order of 1.5 km per month; they suggest that mass movement rather than vertical diffusion is the dominant mechanism. Volchok,[82] besides obtaining residence times of debris as reported earlier, developed a worldwide fallout model for high-altitude explosions, accepting the stratospheric motion interpretations of his colleagues. Several years later in 1969, using [102]Rh, [109]Cd, and [238]Pu as well as some fission-product data, List and Telegadas[77] concluded: "The tracer data indicate a summer-to-winter hemisphere flow above about 37 km and a mean descending motion in the winter stratosphere between 25° and about 70°. Ascending motion occurs near the equatorial tropopause and in the lower winter stratosphere poleward of 70°. Virtually the entire summer stratosphere and the winter stratosphere equatorward of 25° between 18 and 25 km is dominated by mixing processes with no evidence of organized circulations in the meridional plane." They deduced a schematic representation of stratospheric circulation which is reproduced here (Fig. 8). List and Telegadas state that the tracer data should not be ignored in the process of constructing models of the large-scale circulation features of the stratosphere from other considerations. Later, Machta, Telegadas, and List[87] provided further support for this statement.

Finally, Krey and Krajewski[88] developed "a semi-empirical box model of atmospheric transport that permits the calculation of stratospheric inventories, surface air concentrations, and deposition of debris injected into the stratosphere, mesosphere, or higher levels. The model divides the atmosphere of each hemisphere into three compartments; the atmosphere above 21 km, the stratosphere below 21 km, and the troposphere. The transfer between compartments follows first-order kinetics, although the season and height of injection regulate the onset of the transfer. The model adequately computed the fallout parameters of the specific injections of [102]Rh, [109]Cd, [238]Pu, and ^{90}Sr from the 1961-1962 tests, and of ^{90}Sr from the sixth Chinese nuclear test in June 1967. It also predicted the 1969 fallout from the recent atmospheric tests."

Peterson's[85] half-residence time for [238]Pu resulting from the burnup in 1964 of a nuclear-powered satellite at 46 km above the Indian Ocean is 3.5 years. Note that these debris, in contrast to those of high-altitude explosions, were not subjected to an upwards thrust.

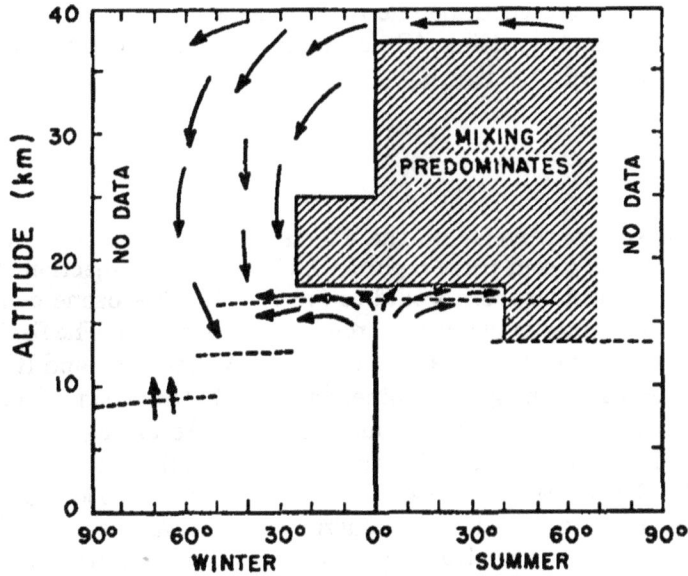

Fig. 8.

Schematic representation of the stratospheric circulation as deduced from radioactive tracer data. *

It remains to be seen whether or not observations of radioactive tracers and their interpretations have made an impact on the science of global atmospheric and stratospheric physics. In 1968, fallout samples from high-altitude explosions became too small to be of further use, and the number of pertinent publications decreased after 1972. The US National Report to the International Union of Geodesy and Geophysics for the years 1971-1974[89] does not seem to take cognizance of the earlier work—probably because the report concentrated on research done during a later period. On the other hand, Reiter, in "Atmospheric Transport Processes," an AEC publication,[90] refers extensively to radioactive-tracer observations when discussing strong vertical mixing processes of stratospheric air via the jetstream into the troposphere. Very recently the same author[91] utilizes the worldwide observations of the motion of the tracers in a more detailed review of the exchange of air masses between troposphere and stratosphere.

Carbon-14

All nuclear explosions produce amounts of ^{14}C. The 1969 UN report on the effects of atomic radiation[92] provides data on the stratospheric and tropospheric content in both hemispheres. While the concentrations in the northern hemisphere have gradually decreased from 1963 to 1967, they remained essentially constant in the southern hemisphere as a consequence of interhemispheric mixing. In 1967, the tropospheric content of explosion-produced ^{14}C was about 65% of the natural level. Naturally produced ^{14}C originates also mainly in the stratosphere. The stratospheric residence time of bomb-produced ^{14}C is quoted as two to five

Reference 83 also shows circulation patterns for the troposphere and the lower stratosphere derived from low-altitude megaton-size explosions. Cloud tops and bases as functions of yield and latitude.

40

years.[83] A fraction of this was produced by high-altitude explosions. (In the same context, the residence time of ^{90}Sr is reported to be one to two years!)

I understand that researchers doing ^{14}C dating need to make certain corrections in their analyses for bomb-produced ^{14}C (Ref. 93).

Bromine

In view of the current controversy on the effects of halogen gases on natural ozone and because of highly confused recent press reports which claim that injection of a few kilograms of bromine gas into the stratosphere would seriously affect the ozone concentration, John Zinn and I looked into the matter of bromine production by fission. The fission product chain yield for stable ^{81}Br is of the order of 3×10^{-3} (courtesy W. Sedlacek and K. Wolfsberg); thus 1 Mt of fission produces about 65 g of bromine, and 200 Mt (the estimated fission yield of all the 1961-1962 atmospheric tests) yields about 13 kg ^{81}Br. According to A. L. Lazrus et al.[94] and personal communication by W. Sedlacek, the mass mixing ratio of bromine/air in 1974 was about 8×10^{-12} g Br per gram air at altitudes of the order of 25 km, or a total worldwide content of 10^6 kg Br in a 5-km-thick stratospheric volume near the peak of the ozone layer. Thus, stratospheric injection of nuclear-explosion-produced bromine or of any other type of kilogram-size bromine injection has no additional effect whatsoever. Fission-produced iodine is roughly 10 times as abundant as bromine, but its injection is also insignificant compared with the current worldwide inventory of bromine and chlorine; furthermore, while the iodine reaction rates with O_3 and its catalytic effects do not seem to be known well at this time, they are probably of the same order of magnitude as those of other halogens.

Local Fallout

No significant local fallout or induced activity was observed on any of the high-altitude events.

41

CHAPTER X

HOLE IN THE OZONE LAYER AFTER TEAK AND ORANGE?

In late 1957 and early 1958, the question was raised as to whether or not the ultraviolet emissions from the Teak and Orange events would "burn a hole" into the natural ozone layer. The pre-event discussions[95] were inconclusive. It was recognized that the ultraviolet radiation in the photon energy range from 4 to 6.5 eV (~3 000 Å to 2 000 Å) would be absorbed by O_3, leading to dissociation; however, absorption of still shorter uv radiation in the range from 6.5 to 11 eV (~2 000 Å to 1 000 Å) in the Schumann-Runge continuum would lead to dissociation of O_2 and subsequent formation of ozone. The general feeling was that destruction and formation would balance each other. This feeling was strengthened by the fact[96] that significant amounts of ozone are produced in sea-level explosions. Furthermore, it was argued that even in case of complete destruction of the ozone layer over an area with radius 50 km, the ozone loss would amount to only 2×10^{-5} of the global inventory. The "hole" would be closed promptly by bomb-produced turbulence and ambient motions in the atmosphere.

After the events, little attention was paid to this particular problem, evidently because no spectacular or unusual observations were made (because of lack of evidence one way or the other). A recent re-inspection of spectra taken by NRL with quartz optics showed, for both Teak and Orange, the usual cutoff near 3 000 Å.

After the event, the Teak fireball uv outputs were calculated at LASL by Skumanich, using the best air opacities available at that time. The calculations show that, during the main radiative phase, about equal amounts of thermal energy would be captured by ozone and by the Schumann-Runge continuum of O_2. However, the ozone-destroying process and the ozone formation by dissociated molecular oxygen have different altitude dependence. A precise treatment of this problem would be a desirable and certainly possible task. It would require application of fireball phenomenology, energy deposition, and air chemistry codes. In the absence of such calculations, it still appears that destruction and formation balance each other. The NO_x formed inside the fireball was carried to altitudes in excess of 100 km and was probably not very effective in attacking the natural ozone layer. No pertinent calculations were done for Orange. The medium-yield Bluegill event, also fired above the ozone layer, was thoroughly analyzed. No emission was observed below 3 000 Å. Calculations of the uv-integrated power at wavelengths below 3 000 Å did not permit definite conclusions. In any case, it appears that the US high-altitude tests with a total yield of the order of 10 Mt had very little (if any) effect on the natural ozone layer.

This is understandable in view of the results of numerous recent theoretical studies relating variations of the natural ozone in the years 1961-1964 to the massive nuclear tests of this period. Much of the NO_x produced by a total energy release of about 340 Mt, mainly from the Russian test series in Novaya Zemlya, was carried close to and into the ozone layer. The precise effects are still under dispute; they are partially obscured by natural fluctuations. A temporary depletion of about 6% is the highest number which has appeared in the literature;[97] however, other investigators[98] feel that the fluctuations observed during the critical period lie within the probable error of available ozone measurements.

CHAPTER XI

CONCERN ABOUT ENVIRONMENTAL EFFECTS OF STARFISH-TYPE EXPLOSIONS: BRITISH AND US-NASA REACTIONS IN 1962. SCIENTIFIC VALUES IN RETROSPECT.

In May 1962, Sir Bernard Lovell, Director of the Radio Astronomy Laboratories, Jodrell Bank, delivered an address before the British Institute of Strategic Studies on the "Challenge of Space Research."[99] After viewing the scientific dividends of space research in the brief period since 1958, he deplored the inability of British participation in major space-based activities and suggested "a new outlook and a new budget on the biggest possible scale." He praised the realization of the new situation by the USA, "because it is evident, that the battle between East and West is seen by the USSR as a conflict in the field of science and technology. This issue has been joined by the Americans...." However, he goes on to say:

"May I conclude by saying that in spite of this enthusiasm which I display and this optimism for the future of scientific research I must confess that my belief in the inevitability of prograss has been very considerably undermined during this past year by the realization that some of the American and probably some of the Russian space activities are not being guided by the purest of scientific motives. I refer of course to the military programme by the U.S. Air Force for the orbiting of "Project Needles (West Ford)" around the Earth and more recently the proposed explosion of a megaton nuclear weapon in the region of the Van Allen belts.... These subjects are unusually controversial, and the only point I would make now is to emphasize the extreme importance of carrying out such projects only by agreement of the International Scientific Unions. If I might end on this rather sad note I do beg the Americans to use their influence to the utmost to make quite sure that such projects are carried out only by international agreement and particularly within the framework of the resolution of the International Astronomical Union which was phrased only a few months ago in California. If this is not done then the United States may bear the awful responsibility of having started a chain of events leading to the militarization of space and the destruction of astronomy on Earth.

"Note added in proof by the author. This address was delivered to the Institute of Strategic Studies nearly four months ago. In this time there have been further space activities which underline the anxiety expressed in the last paragraph of my address....The United States exploded the megaton bomb outside the atmosphere and have thereby enormously confused the study of the natural radiation belts by setting up a new long-duration zone of trapped particles. A few more explosions of this type for military purposes by other of the Powers will obviously add so much artificially trapped material to the radiation zones that the investigation of the natural effects will have to be abandoned before we know their true nature or origin....Finally, a highly successful communication satellite, Telstar, has been placed in orbit and thereby encouraged the commercial as well as the military communication interests in space. The anxiety expressed about 'Project Needles' must be paralleled by the anticipation that many Telstar or Echo balloon satellites will have a similar detrimental effect on earthbound astronomy and

45

radioastronomy. The need for international agreement about the use of space and the control of launchings, either of rockets or space-vehicles into it, has become a matter of the utmost urgency."

Similar views have been expressed by several editorials of "Nature."[100,101]

In the United States, the reaction to the Starfish event was mixed. Many members of the NASA staff expressed great concern about possible interference of the artificial radiation belts with the space program. Their anxiety was enhanced by the fact that two or three satellites were put out of commission by the artificial electron fluxes, thus terminating their missions, and by the premature claims (September 1962) of the Bell space scientists that essentially all fission beta rays were injected into the belts; (as it turned out, it was ~5%, not "more than 50%," of the pertinent activity). In any case, Dr. Webb, the NASA administrator at that time, prevailed upon Dr. Jerome Wiesner, the Chief Scientific Advisor to the President, and reportedly also directly upon President Kennedy to have future nuclear space experiments restricted to lower altitudes. This, in my personal opinion, highly emotional response led unfortunately to the cancellation of the low-yield Uracca event, which was to be exploded at an altitude of 1 300 km as proposed by LASL. The event, as planned, would have added less than 1% to the inventory of the artificial belts but would have increased our knowledge of near-space physics significantly.

Another critical response, though of a different type, might be worth recording. Nawrocki[69] in discussing the physics of magnetic disturbances (see Chapter VI) writes:

"The ambiguity concerning interpretation of the magnetic disturbance is just one indication of how poorly conceived and instrumented were the Argus experiments. Perhaps all that was achieved was the substantiation of the existence of the terrestrial B-field. Contrary to Christofilos[27] and others, the high-altitude explosion is not a good tool for investigating the atmosphere. Rather one must know the atmosphere extremely well to interpret the highly complex interaction of the bomb and the perturbed environment."

This statement was made in 1961. Nawrocki was right at least in one respect, namely that Argus was poorly instrumented.

As time went by, the emotions were replaced by genuine scientific curiosity and a thorough, high-class analysis of the many unusual phenomena observed as a consequence of the high-altitude events. Literally many hundreds of publications in scientific magazines (such as the "Journal of Geophysical Research" and the "New Zealand Journal of Geology and Geophysics;" in the "Proceedings of the National Academy of Sciences;" in "Cospar Publications" and in "Nature" (sic), etc.; and in unclassified documents of the AEC, of AEC Laboratories, and of DOD agencies and their contractors) bear witness to the stimulating questions which these events raised and frequently answered. Studies were initiated which otherwise would not have been conducted. True, some of the studies suffered from the classification of the source outputs. Still, now in retrospect, I believe that the advances in our knowledge in many fields of physics did outweigh the disadvantages. After all, the results of many large-scale experiments are derived from the modification or the simulation of natural processes.

It is worthwhile to quote here a few published postevent positive reactions to several phases of the observations and to list several—of many—rather basic scientific rewards which were a direct consequence of the high-altitude experiences.

In a United Nations document published in 1972 we read: "Observations of radioactive tracers have contributed greatly to the understanding of air movement within the stratosphere...."[102]

In a textbook on *Particles in the Atmosphere and Space*[103] we read that radioactive "particles behave almost as though they are molecules, and they enter the troposphere only as the stratospheric air enters the troposphere. Because of this behavior and because radioactive particles are readily identified as such, delayed fallout serves as an excellent tracer for air masses."

46

In another publication[56] it is stated: "Geomagnetic and other perturbations produced by such explosions can lead to an understanding of certain geophysical processes. For example, explosions like this [Starfish] provide an opportunity to measure the time delay of the hydromagnetic waves they generate, and thus they furnish evidences about the propagation mechanism of the waves."

In an otherwise classified document, H. A. Bethe[104] wrote in 1957, before the Teak event, as follows:

"The many deviations from equilibrium which are characteristic of high-altitude shots make it very difficult to make definite predictions on either hydrodynamic or optical phenomena....All I can hope to do is indicate the scale of the phenomena, not the details. This makes it more interesting to make observations on this test. In fact, the test will constitute a beautiful laboratory for the study of the properties of air in large quantity and at very low density. It is regrettable that the test can not be planned with greater leisure and instrumented more fully to make use of this important opportunity to study the properties of the high atmosphere."

This remark by Bethe and the analysis of data established the need for better, more precise reaction rates for atmospheric constituents. In the subsequent years, an extensive research program conducted mainly at academic institutions under the Defense Atomic Support Agency (DASA) and later the Defense Nuclear Agency (DNA) sponsorship resulted in the publication of a "Reaction Rate Handbook"[105] which is frequently being updated. It is an essential source of information for atmospheric researchers.

The application of the experiences gained in 1962 during the AEC's rocket-borne diagnostic experiments to the Vela Satellite program provided another dividend.[106] This program combined the objective of detecting possible clandestine nuclear explosions in space with fruitful basic magnetospheric research.[107,108]

Finally, the study of the physics of the interaction of debris with low-density air (the so-called coupling processes) provided the impetus for the intensification of several phases of modern plasma physics as demonstrated by Longmire.[109]

ACKNOWLEDGMENTS

This work was done under the auspices of the Nevada Operations Office of the US Energy Research and Development Administration, with Ross L. Kinnaman as the Manager of the NNTRP Environmental Assessment Program. Thanks are due to members of the reference section of the LASL Library for their valuable assistance. Eugene J. Baumann of the Radio Physics Laboratory of the Stanford Research Institute provided many references for the section on communication degradation. Photographs are by Elbert Bennett, William Regan, and Roy Stone, all of LASL, and by members of the EG&G teams. Among many of my LASL colleagues who have read the draft manuscript, I am particularly indebted to Robert Brownlee, Marvin Hoffman, John Malik, Carson Mark, and John Zinn for their helpful comments.

48

REFERENCES

1. S. Glasstone, *The Effects of Nuclear Weapons* (US Atomic Energy Commission, April 1962). Available through US Government Printing Office.

2. J. Zinn, Los Alamos Scientific Laboratory, personal communication, March 1965.

3. S. L. Severin, A. V. Alder, N. L. Newton, and J. F. Culver, "Photostress and Flash Blindness in Aerospace Operations," USAF School of Aerospace Medicine, Brooks AFB, TX, SAM-TDR-63-67 (September 1963).

4. D. W. Williams and B. C. Duggar, "Review of Research on Flash Blindness, Chorioretinal Burns, Countermeasures, and Related Topics," Defense Atomic Support Agency report DASA-1576 (1965).

5. R. G. Allen Jr., W. R. Bruce, K. R. Kay, L. K. Morrison, R. A. Neish, C. A. Polaski, and R. A. Richards, "Research on Ocular Effects Produced by Thermal Radiation," Technology, Inc. report prepared for USAF School of Aerospace Medicine, Aerospace Medical Division, Brooks AFB, TX, Appendix B (July 1967).

R. G. Allen Jr., D. J. Isgitt, D. E. Jungbauer, J. H. Tips Jr., P. W. Wilson Jr., and T. J. White, "The Calculation of Retinal Burn and Flashblindness Safe Separation Distances," USAF School of Aerospace Medicine, Aerospace Medical Division, Brooks AFB, TX, SAM-TR-68-106 (September 1968).

6. W. T. Ham Jr., R. C. Williams, W. J. Geeraets, R. S. Ruffin, and H. A. Mueller, "Optical Masers (Lasers)," Acta Opthalmol., Suppl. **76**, 60 (1963).

7. N. D. Miller and T. J. White, "Evaluation of Eye Hazards from Nuclear Detonations I. Retinal Burns and Flash Blindness," USAF School of Aerospace Medicine, Aerospace Medical Division, Brooks AFB, TX (November 1969).

8. H. Hoerlin, A. Skumanich, and D. R. Westervelt, Los Alamos Scientific Laboratory, personal communication, August 1959.

9. H. L. Mayer and F. Ritchey, "Eyeburn Damage Calculations for an Exospheric Nuclear Event," J. Opt. Soc. Am. **54**, 678 (1964).

10. R. D. Cowan, "Calculation of Retinal Dose Due to Visible Radiation From Nuclear Explosions," Los Alamos Scientific Laboratory report LA-3204-MS (March 1965).

11. J. Zinn, R. C. Hyer, and C. A. Forest, "Eyeburn Thresholds," Los Alamos Scientific Laboratory report LA-4651 (May 1971).

49

12. W. F. Culver, N. L. Newton, R. Penner, and R. W. Neidlinger, "Human Chorioretinal Burns Following High Altitude Nuclear Detonations," Aerosp. Med. **35**, 1217 (December 1964).

13. H. A. Bethe, Los Alamos Scientific Laboratory consultant, personal communication, August 1958.

14. A. Skumanich, Los Alamos Scientific Laboratory, personal communication, January 1958.

15. W. Ogle, Los Alamos Scientific Laboratory, personal communication, March 1968.

16. H. Paul Williams, "The Effect of High-Altitude Nuclear Explosions on Radio Communications," IRE Trans. Mil. Electron. (Inst. Radio Eng., New York), 326 (October 1962).

17. W. F. Utlaut, "Ionospheric Effects Due to Nuclear Explosions," Boulder Laboratories, National Bureau of Standards report 6050 (April 30, 1959).

18. G. C. Andrew, "Some Observations of MF and HF Radio Signals After Mid-Pacific High-Altitude Nuclear Explosions," New Zealand J. Geol. Geophys. **5**, 988 (December 1962).

19. T. Obayashi, S. C. Coroniti, and E. T. Pierce, "Geophysical Effects of High-Altitude Nuclear Explosions," Nature **183**, 1476 (May 23, 1959).

20. D. Davidson, "Nuclear Burst Effects on Long-Distance High-Frequency Circuits," J. Geophys. Res. **68**, 331 (January 1, 1963).

21. R. G. D'Arcy and Stirling A. Colgate, "Measurements at the Southern Conjugate Region of the Fission Debris from the Starfish Nuclear Detonation," J. Geophys. Res. **70**, 3147 (July 1, 1965).

22. A. J. Zmuda, B. W. Shaw, and C. R. Haave, "Very Low Frequency Disturbances and the High-Altitude Nuclear Explosion of July 9, 1962," J. Geophys. Res. **68**, 745 (February 1, 1963).

23. R. P. Basler, R. B. Dyce, and H. Leinbach, "High-Latitude Ionization Associated with the July 9 Explosion," J. Geophys. Res. **68**, 741 (February 1, 1963).

24. James J. Griffin, "Re-Evaluation of Optimal Parameters for Post-Fission Beta Decay," Los Alamos Scientific Laboratory report LA-2811, Addendum II (June 1964).

25. Herman Hoerlin, "Artificial Aurora and Upper Atmospheric Shock Produced by Teak," Los Alamos Scientific Laboratory report LAMS-2536 (June 1961).

26. N. C. Christofilos, "The Argus Experiment," J. Geophys. Res. **64**, 869 (August 1959).

27. N. C. Christofilos, "Sources of Artificial Radiation Belts," in *Radiation Trapped in the Earth's Magnetic Field,* B. M. McCormack, Ed. (D. Reidel Publ. Co., Dordrecht-Holland, 1966), pp. 565-574.

50

28. James A. Van Allen, Carl E. McIlwain, and George H. Ludwig, "Satellite Observations of Electrons Artificially Injected into the Geomagnetic Field," J. Geophys. Res. **64**, 877 (August 1959).

29. Lew Allen Jr., James L. Beavers II, William A. Whitaker, Jasper A. Welch Jr., and Roddy B. Walton, "Project Jason Measurement of Trapped Electrons From a Nuclear Device by Sounding Rockets," J. Geophys. Res. **64**, 893 (August 1959).

30. W. N. Hess, *The Radiation Belt and Magnetosphere* (Blaisdell Publishing Co., Waltham, MA, 1968).

31. J. A. Van Allen, "Spatial Distribution and Time Decay of the Nuclear Intensities of Geomagnetically Trapped Electrons From the High Altitude Nuclear Burst of July 1962," in *Radiation Trapped in the Earth's Magnetic Field,* B. M. McCormack, Ed. (D. Reidel Publ. Co., Dordrecht-Holland, 1966), pp. 575-592.

32. J. B. Cladis, G. T. Davidson, and L. N. Newkirk, "The Trapped Radiation Handbook," Defense Nuclear Agency report DNA-2524H (December 1971).

a. J. Billingham, "Status Report on the Space Radiation Effects on the Apollo Mission," in *Second Symposium on Protection Against Radiations in Space*, NASA SP-71, A. Reetz Jr., Ed., Scientific and Technical Information Division, National Aeronautics and Space Administration, Washington, DC, 1965.

b. W. H. Langham, Ed., *Radiobiological Factors in Manned Space Flight*, National Academy of Sciences and National Research Council Publication 1487 (Washington, DC, 1967), pp. 14-20.

33. R. W. Kilb, "Analysis of Argus III Photographic Data (U)," Unclassified parts of Mission Research Corporation report MRC-R-112 (January 1974).

34. R. W. Kilb, "Analysis of Argus II All-Sky Photographs (U)," Unclassified parts of Mission Research Corporation report MRC-R-176 (March 1975).

35. Symposium on Scientific Effects of Artificially Introduced Radiations at High Altitudes held April 29, 1959, at the National Academy of Sciences, Washington, DC. Papers published in J. Geophys. Res. **64**, 865 (August 1959) and in Proc. Nat. Acad. Sci. USA, **45**, 1141 (August 15, 1959).

36. G. L. Johnson and R. B. Dyce, "A Study of Explorer IV Records in the Pacific Area," Stanford Research Institute report FO-0-111 (July 1960).

37. B. J. O'Brien, C. D. Laughlin, and J. A. Van Allen, "Geomagnetically Trapped Radiation Produced by a High-Altitude Nuclear Explosion on July 9, 1962," Nature **195**, 939 (September 8, 1962).

38. W. L. Brown, W. N. Hess, and J. A. Van Allen, "Collected Papers on the Artificial Radiation Belt from the July 9, 1962, Nuclear Detonation," J. Geophys. Res. **68**, 605 (February 1, 1963).

39. J. A. Van Allen, L. A. Frank, and B. J. O'Brien, "Satellite Observations of the Artificial Radiation Belt of July 1962," J. Geophys. Res. **68**, 619 (February 1, 1963).

40. W. N. Hess, "The Artificial Radiation Belt Made on July 9, 1962," J. Geophys. Res. **68**, 667 (February 1, 1963). Actually, Hess refers to Brown and Gabbe, J. Geophys. Res. **68**, 607 (February 1, 1963). Brown and Gabbe's numbers of the total inventory are a factor of three lower than Hess', i.e., more like 7×10^{25} electrons.

41. S. A. Colgate, "Energetic Electrons From Shock Heating in the Exosphere," in *Radiation Trapped in the Earth's Magnetic Field,* B. M. McCormack, Ed. (D. Reidel Publ. Co., Dordrecht-Holland, 1966), p. 693.

42. J. J. Griffin, "Beta Decays and Delayed Gammas From Fission Fragments," Los Alamos Scientific Laboratory report LA-2811 (February 1963).

43. J. Zinn, H. Hoerlin, and A. G. Petschek, "The Motion of Bomb Debris Following the Starfish Test," in *Radiation Trapped in the Earth's Magnetic Field*, B. M. McCormack, Ed. (D. Reidel Publ. Co., Dordrecht-Holland, 1966), pp. 671-692.

44. H. Hoerlin, "Air Fluorescence Excited by High-Altitude Nuclear Explosions," Los Alamos Scientific Laboratory report LA-3417-MS (January 1966).

45. A. C. Durney, H. Elliott, R. J. Hynds, and J. J. Quenby, "Satellite Observations of the Energetic Particle Flux Produced by the High-Altitude Nuclear Explosion of July 9, 1962," Nature **195**, 1245 (September 29, 1962).

46. G. F. Pieper and D. J. Williams, "Traac Observations of the Artificial Radiation Belt from the July 9, 1962 Nuclear Detonation," J. Geophys. Res. **68**, 635 (February 1, 1963).

47. Yu. I. Galperin and A. D. Bolyunova, "Study of the Drastic Changes of the Radiation in the Upper Atmosphere in July, 1962," in *Space Research V* (North-Holland Publ. Co., Amsterdam, 1965), p. 446.

48. V. I. Krassovsky, Yu. I. Galperin, V. V. Temnyy, T. M. Mulyarchik, N. V. Dzhordzhio, M. Ya. Marov, A. D. Bolyunova, O. L. Vaysberg, B. P. Potapov, and M. L. Bragin, "Some Characteristics of Geoactive Particles," Geomagn. Aeron. **3**, 401 (1963).

49. D. A. Adams and B. W. Mar, "Predictions and Observations for Space Flight," in *Radiation Trapped in the Earth's Magnetic Field*, B. M. McCormack, Ed. (D. Reidel Publ. Co., Dordrecht-Holland, 1966), p. 817.

50. J. Hazel, "Radiation Hazards and Manned Space Flight," Aerosp. Med. **35**, 436 (May 1964).

51. B. M. McCormack, Ed., Session VI, "Artificial Injected Radiation," and Session IX, "Radiation Doses Received by Manned Flight in the Trapped Radiation Belts," in *Radiation Trapped in the Earth's Magnetic Field*. (D. Reidel Publ. Co., Dordrecht-Holland, 1966), pp. 565-702 and 817-886.

52

52. R. B. Dyce and S. Horowitz, "Measurements of Synchrotron Radiation at Central Pacific Sites," J. Geophys. Res. **68**, 713 (February 1963).

53. A. M. Peterson and G. L. Hower, "Synchrotron Radiation From High-Energy Electrons," J. Geophys. Res. **68**, 723 (February 1, 1963).

54. G. R. Ochs, D. T. Farley Jr., K. L. Bowles, and P. Bandyopadhay, "Observations of Synchrotron Radio Noise at the Magnetic Equator Following the High-Altitude Nuclear Explosion of July 9, 1962," J. Geophys. Res. **68**, 701 (February 1, 1963).

55. J. M. Hornby, "The Effects on Radio-Astronomy of the July 9, 1962, High-Altitude Nuclear Explosion," J. Geophys. Res. **69**, 2737 (July 1, 1964).

56. M. Casaverde, A. Giesecke, and Robert Cohen, "Effects of the Nuclear Explosion Over Johnston Island Observed in Peru on July 9, 1962," J. Geophys. Res. **68**, 2603 (May 1, 1963).

57. R. L. Kovach, and A. Ben-Menahem, "Analysis of Geomagnetic Micropulsations Due to High-Altitude Nuclear Explosions," J. Geophys. Res. **71**, 1427 (March 1, 1966).

58. J. R. Davis and J. M. Headrick, "A Comparison of High-Altitude Nuclear Explosion Effects in the E Layer with Variations in Geomagnetic Field Strength," J. Geophys. Res. **69**, 911 (March 1, 1964).

59. P. J. Nawrocki and R. Papa, "Transmission of Electromagnetic Waves," in *Atmospheric Processes* (Prentice-Hall, Inc., Englewood Cliffs, New Jersey, 1963), Chap. 10, pp. 19-21.

60. Paul L. Hartman, "New Measurements of the Fluorescence Efficiency of Air Under Electron Bombardment," Planet. Space Sci. **16**, 1315 (1968). Also Los Alamos Scientific Laboratory report LA-3793 (January 1968).

61. Robert O'Neil and Gilbert Davidson, "The Fluorescence of Air and Nitrogen Excited by Energetic Electrons," American Science and Engineering report ASE-1602, AFCRL-67-0277 (January 1968).

62. Kenneth B. Mitchell, "Fluorescence Efficiencies and Collisional Deactivation Rates of N_2 and N_2^+ Bands Excited by Soft X-Rays," J. Chem. Phys. **53**, 1795 (September 1970).

63. D. R. Westervelt, E. W. Bennett, and A. Skumanich, "Air Fluorescence Excited by Gamma Rays and X-Rays," Proc. 4th Intern. Conf. Ionization Phenomena in Gases," (North-Holland Publ. Co., Amsterdam, 1960), Vol. 1, p. 225.

64. J. W. Chamberlain, *Physics of the Aurora and Airglow* (Academic Press, New York, 1961), p. 197.

65. Sadami Matsushita, "On Artificial Geomagnetic and Ionospheric Storms Associated with High-Altitude Explosions," J. Geophys. Res. **64**, 1149 (September 1959).

53

66. J. B. Gregory, "New Zealand Observations of the High-Altitude Explosion of July 9 at Johnston Island," Nature **196**, 508 (November 10, 1962).

67. Special Nuclear Explosion Issue, New Zealand J. Geol. Geophys. **5**, 916-1018 (December 1962).

68. H. Argo, H. Hoerlin, C. Longmire, A. Petschek, and A. Skumanich, "Nuclear Explosions in Space," Proc. 2nd Plowshare Symp., San Francisco, May 1959, University of California report UCRL-5679 (1959).

69. D. R. Westervelt and H. Hoerlin, "The Los Alamos Air Fluorescence Detection System," Proc. IEEE, p. 2068 (December 1965).

70. A news item in EOS, Trans. Am. Geophys. Union, **54**, 687 (July 1973), referring to work by W. Orr Roberts of the University Corporation for Atmospheric Research and by Roger H. Olson, Research Meteorologist with National Oceanic and Atmospheric Administration.

71. W. Orr Roberts and R. J. Olson, "Geomagnetic Storms and Wintertime 300-mb Trough Development in the North Pacific-North America Area," J. Atmos. Sci. **30**, 135 (January 1973).

72. J. M. Wilcox, P. H. Scherrer, L. Svalgaard, W. Orr Roberts, and R. H. Olson, "Solar Magnetic Sector Structure: Relation to Circulation of the Earth's Atmosphere," Science **180**, 185 (April 1973).

W. Orr Roberts and R. H. Olson, "New Evidence for Effects of Variable Solar Corpuscular Emission on the Weather," Rev. Geophys. Space Phys. **11**, 731 (August 1973).

J. M. Wilcox, P. H. Scherrer, L. Svalgaard, W. Orr Roberts, R. H. Olson, and R. L. Jenne, "Influence of Solar Magnetic Structure on Terrestrial Atmospheric Vorticity," J. Atmos. Sci. **31**, 581 (March 1974).

J. M. Wilcox, L. Svalgaard, and P. H. Scherrer, "Seasonal Variation and Magnitude of the Solar Sector Structure-Atmospheric Vorticity Effect," Nature **255**, 539 (June 12, 1975).

R. H. Olson, W. Orr Roberts, and C. S. Zerefos, "Short Term Relationships Between Solar Flares, Geomagnetic Storms and Tropospheric Vorticity Patterns," Nature **257**, 133 (September 11, 1975).

J. M. Wilcox, "Solar Structure and Terrestrial Weather," Science **192**, 745 (May 1976).

73. W. F. Utlaut and R. Cohen, "Modifying the Ionosphere with Intense Radio Waves," Science **174**, 245 (1971).

F. W. Perkins and S. H. Francis, "Artificial Production of Traveling Ionospheric Disturbances and Large Scale Atmospheric Motion," J. Geophys. Res. **79**, 3879 (September 1, 1974).

David S. Evans, "High-Velocity Gas Releases as a Method of Perturbing the Upper Atmosphere," J. Geophys. Res. **79**, 3882 (September 1, 1974).

54

74. M. I. Kalkstein, "Rhodium-102 High-Altitude Tracer Experiment," Science 137, 645 (August 31, 1962).

75. G. A. Cowan, Remarks prepared for "Seminar on Cadmium Isotopes from Starfish" at HASL (December 6, 1963).

N. A. Hallden and L. P. Salter, "Emitted Radiations from Cadmium-109 and Cadmium-113m," USAEC, New York Operations Office, Health and Safety Laboratory report HASL-142 (January 1, 1964), p. 299.

L. P. Salter, "Note on the Detectability of Cadmium Isotopes from Starfish in 1964 Ground Level Samples," HASL-142 (January 1, 1964), p. 303.

I. J. Russell and R. V. Griffith, "The Production of Cd^{109} and Cd^{113m} in a Space Nuclear Explosion," HASL-142 (January 1, 1964), p. 306.

76. K. Telegadas and R. J. List, "Global History of the 1958 Nuclear Debris and its Meteorological Implications," J. Geophys. Res. 69, 4741 (November 15, 1964).

77. R. J. List and K. Telegadas, "Using Radioactive Tracers to Develop a Model of the Circulation of the Stratosphere," J. Atmos. Sci. 26, 1128 (November 1969).

78. L. Machta, R. J. List, and K. Telegadas, "Inventories of Selected Long-lived Radioisotopes Produced During Nuclear Testing," USAEC, New York Operations Office, Health and Safety Laboratory report HASL-142 (January 1, 1964), pp. 255, 270.

79. K. Telegadas, "The Seasonal Stratospheric Distribution of Cadmium-109, Plutonium-238 and Strontium-90," USAEC, New York Operations Office, Health and Safety Laboratory report HASL-184 (January 1, 1968), pp. I-53—I-85.

80. R. J. List, L. P. Salter, and K. Telegadas, "Radioactive Debris as a Tracer for Investigating Stratospheric Motions," Tellus 18, 345 (1966).

81. O. I. Leipunskii, J. E. Konstantinov, G. A. Fedorov, and O. G. Scotnikova, "Mean Residence Time of Radioactive Aerosols in the Upper Layers of the Atmosphere Based on Fallout of High-Altitude Tracers," J. Geophys. Res. 75, 3569 (June 20, 1970).

82. H. L. Volchok, "The Anticipated Distribution of Cd-109 and Pu-238 (from SNAP-9A) Based Upon the Rh-102 Tracer Experiments," USAEC, New York Operations Office, Health and Safety Laboratory report HASL-165 (January 1, 1966), p. 312.

83. "Ionizing Radiation: Levels and Effects," a report of the United Nations Scientific Committee on the Effects of Atomic Radiation to the General Assembly (A United Nations Publication, New York, 1972), p. 40.

84. F. Fabian, W. F. Libby, and C. E. Palmer, "Stratospheric Residence Time and Interhemispheric Mixing of Strontium 90 from Fallout in Rain," J. Geophys. Res. 73, 3611 (June 15, 1968).

55

85.   K. R. Peterson, "An Empirical Model for Estimating World-Wide Deposition from Atmospheric Nuclear Detonations," Health Phys. **18**, 357 (1970).

86.   A. K. Stebbins III, "Special Report on High Altitude Sampling Program," Defense Atomic Support Agency report DASA-532B (June 1, 1960).

A. K. Stebbins III, "Second Special Report on High Altitude Sampling Program," Defense Atomic Support Agency report DASA-539B (August 1, 1961).

87.   L. Machta, K. Telegadas, and R. J. List, "The Slope of Surfaces of Maximum Tracer Concentration in the Lower Stratosphere," J. Geophys. Res. **75**, 2279 (April 20, 1970).

88.   P. W. Krey and B. Krajewski, "HASL Model of Atmospheric Transport," USAEC, New York Operations Office, Health and Safety Laboratory report HASL-215 (September 1969).

P. W. Krey and B. Krajewski, "Comparison of Atmospheric Transport Model Calculations with Observations of Radioactive Debris," J. Geophys. Res. **75**, 2901 (May 20, 1970).

89.   US National Report to the International Union of Geodesy and Geophysics, 1971-1974, Rev. Geophys. Space Phys., Vol. 13 (July 1975).

90.   E. R. Reiter, "Atmospheric Transport Processes, Part 3," AEC Critical Review Series, USAEC report TID-25731 (1972).

91.   E. R. Reiter, "Atmospheric and Tropospheric Exchange Processes," Rev. Geophys. Space Phys. **13** (1975).

92.   Report of the United Nations Scientific Committee on the Effects of Atomic Radiation, Official Record, Twenty-Fourth Session (A United Nations Publication, New York, 1969), Suppl. 13, p. 14.

93.   Anthony Turkevich, University of Chicago and Los Alamos Scientific Laboratory, personal communication, August 1975.

94.   A. L. Lazrus, B. W. Bandrud, R. N. Woodward, and W. A. Sedlacek, "Stratospheric Halogen Measurements," Geophys. Res. Lett. **2**, 439 (October 1975).

95.   Discussions took place in Washington, DC, in January 1958 with J. Chamberlain (Yerkes Observatory), F. Gilmore (Rand Corporation), H. Steward (NRL), and H. Hoerlin (LASL) participating.

96.   H. E. DeWitt, "A Compilation of Spectroscopic Observations of Air Around Atomic Bomb Explosions," Los Alamos Scientific Laboratory report LAMS-1935 (June 1955).

97.   H. Johnston, G. Whitten, and J. Birks, "The Effect of Nuclear Explosion on Stratospheric Nitric Oxide and Ozone," J. Geophys. Res. **78**, 6107 (September 20, 1973).

98.   E. Bauer and F. R. Gilmore, "Effect of Atmospheric Nuclear Explosions on Total Ozone," Rev. Geophys. Space Phys. **13**, 451 (August 1975).

56

99.   Sir Bernard Lovell, "The Challenge of Space Research," Nature **195**, 935 (September 8, 1962).

100.   Editorial, "The Space Game," Nature **195**, 739 (August 25, 1962).

101.   Editorial, "News and Views, the High-Altitude Nuclear Explosion of July 9, 1962," Nature **195**, 945 (September 8, 1962).

102.   "Ionizing Radiation: Levels and Effects," Pt. 2, "Man-Made Environmental Radiation," a report of the United Nations Scientific Committee on the Effects of Atomic Radiation to the General Assembly (A United Nations Publication, New York, 1972), Vol. I, *Levels*, pp. 39-41.

103.   Richard D. Cadle, *Particles in the Atmosphere and Space* (Reinhold Publ. Corp., New York, 1966), p. 122.

104   H. A. Bethe, Los Alamos Scientific Laboratory consultant, personal communication, 1957.

105.   Defense Nuclear Agency Reaction Rate Handbook, 2nd Edn., DNA, 1948H, March 1972 (distribution unlimited).

106.   S. Singer, "The Vela Satellite Program for Detection of High-Altitude Nuclear Detonations," Proc. IEEE **53**, 1935 (December 1965).

107.   S. J. Bame, "Time-of-Flight Measurements Made with Neutrons from Nuclear Explosions in Space," Bull. Am. Phys. Soc. **9**, 76 (1964).

108.   H. V. Argo, "Introduction to the Scientific Results from the Vela Satellite Program," Trans. Am. Geophys. Union **45**, 623 (1964).

109.   C. L. Longmire, "Notes on Debris-Air-Magnetic Interaction," Rand Corporation Memorandum RM-3386-PR (January 1963).

☆ U.S. GOVERNMENT PRINTING OFFICE 1976—777-034/ 777-297

57

www.ingramcontent.com/pod-product-compliance
Lightning Source LLC
Chambersburg PA
CBHW080524110426

42742CB00017B/3231